Sustainability and Innovation

Sustainability and Innovation

Published Volumes:

Jens Horbach (Ed.)
Indicator Systems for Sustainable Innovation
2005. ISBN 978-3-7908-1553-5

Bernd Wagner, Stefan Enzler (Eds.)
Material Flow Management
2006. ISBN 978-3-7908-1591-7

A. Ahrens, A. Braun, A.v. Gleich, K. Heitmann, L. Lißner
Hazardous Chemicals in Products and Processes
2006. ISBN 978-3-7908-1642-6

Ulrike Grote, Arnab K. Basu, Nancy H. Chau (Eds.)
New Frontiers in Environmental and Social Labeling
2007. ISBN 978-3-7908-1755-3

Marco Lehmann-Waffenschmidt
(Editor)

Innovations Towards Sustainability

Conditions and Consequences

With 38 Figures and 21 Tables

Physica-Verlag
A Springer Company

Professor Dr. Marco Lehmann-Waffenschmidt
Department of Economics
Dresden University of Technology
01062 Dresden
Germany
lewaf@t-online.de

Library of Congress Control Number: 2007932634

ISSN 1860-1030

ISBN 978-3-7908-1649-5 Physica-Verlag Heidelberg New York

Physica-Verlag is a part of Springer Science+Business Media

springer.com

Production: LE-TEX Jelonek, Schmidt & Vöckler GbR, Leipzig
Cover-design: WMX Design GmbH, Heidelberg

SPIN 11549413 88/3180YL - 5 4 3 2 1 0 Printed on acid-free paper

Contents

Foreword

In 1999, the German federal ministry of education and research (BMBF) decided to include two new priorities in its environmental research policy portfolio. One was concerning socio-ecological research aiming at a better understanding of the social dimension of the sustainability triad and the other one was on innovation oriented research aiming at a better understanding how companies and public authorities can influence innovation activities towards the sustainability objectives. The latter priority led to two new programme lines, one targeting at the company level and the other one targeting at the public policy level under the headline of "*framework conditions for innovations towards sustainable development*" (RIW).

The projects funded under the RIW programme were analysing the potential innovation impact of environmental policy measures on the one hand and the sustainability impact of other policies, such as innovation policy, on the other hand. The design of the RIW programme included in addition an international outreach dimension with the organisation of international conferences as well as the establishment of collaboration platforms among the funded projects in order to allow for more general conclusions. The RIW programme followed the BMBF tradition to foster multi- or interdisciplinary cooperation, notably involving academics from economics, policy sciences, and law.

The selected projects were dealing with concrete, often innovative, policy measures or concepts, such as the lead market concept for instance. When looking at the policy topics discussed today, it can be confirmed that the 'right' projects have been chosen. The RIW programme also included a cluster of projects concerning the new chemicals policy of the European Union, the REACH regulation, which recently entered into force. It also included projects concerning policy strategies, addressing the role of public procurement or the potential of the lead market concept, and finally a cluster of projects concerning specific innovation systems such as the electronics market, the CO_2 emission issue in steel production, the water sector, or the recycling of motor vehicles parts. All these topics can be found to date on the policy agenda of both, the national and the European level.

It can be assumed that the results of the projects were taken up by policy makers and acted as a trigger for more academic work in this regard. Current policy declarations - not only by the German environment ministry - calling for a revised understanding of sustainable development as a driver of innovation and not as a barrier underline the relevance of the RIW programme and its outputs. One of its products you are holding in your hands, and I hope you will find it as inspiring and insightful as I do.

I extend my grateful thanks to Marco Lehmann-Waffenschmidt to lead the RIW working group and to put all the material together, and to the authors and contributors who provided new insights into opportunities to use the force of innovation to realise a more sustainable world.

Sevilla, March 2007 Alexander Grablowitz

Preface

In spring 2001 the German "Bundesministerium für Bildung und Forschung" (federal ministry of education and research) launched research grants on the topic „Rahmenbedingungen für Innovationen zum nachhaltigen Wirtschaften" (framework conditions for innovations towards sustainable development, „Förderschwerpunkt RIW", see also `www.riw-netzwerk.de`). It was part of the organizational structure of the "RIW project" to form several working groups out of the accepted project teams with the dedication to gather and organize the knowledge to be developed during the three-years-research grants across the single project teams. As chair of the working group "Innovations and Sustainable Development" I organized several workshops between 2002 and 2005 where we discussed and developed our ideas on this topic. The result of this discourse is now available in the collection of contributions of this volume.

It has been my ambition to reflect the vividness of the intellectual discourse by presenting each of the research contributions together with a critical comment by an expert in the field. Thus, the volume consists of six pairs of a main contribution and a comment and one "triple" (with two main contributions) written by 20 authors. To be sure a volume like this with little more than 200 pages cannot give a truly exhaustive account on the state of the art in a topic so important as the interrelations between innovations and sustainability. But undoubtedly it appears to be worthwile to provide a representative collection of instructive papers on this topic like the present ones combining case studies with principal thoughts.

The volume is divided into three parts: I. New Approaches to Environmental Innovation Policy, II. Innovations and Sustainability, and III. Arrangements in Society and Economy Towards Sustainability. In Part I Christian Lutz, Bernd Meyer, Jan Nill, and Joachim Schleich identify and examine the "Windows of Opportunity for Radical Technological Change in Steel Production and the Influence of CO_2 Taxes" (with a comment by Klaus Rennings), and René Kemp and Stefan Zundel deal with the question of "Environmental Innovation Policy. Is Steering Innovation Processes Possible?" (commented by Frank Beckenbach). Part I is completed by two case studies on the issue of transition management by Joachim Hafkesbrink "Transition Management in the Electronics Industry Innovation System: Systems Innovation towards Sustainability Needs a New Governance Portfolio" and by René Kemp "An Example of a "Managed Transition": The Transformation of the Waste Management Subsystem in the Netherlands (1960-2000)", which are in a "triple format" commonly commented by Klaus Jacob.

Part II "Innovations and Sustainability" contains two contributions: "Leading Innovations to Sustainable Future Markets" by Klaus Fichter and Reinhard Pfriem (commented by Klaus Burmeister), and Niko Paech's paper on "Directional Certainty in Sustainability-Oriented Innovation Management" with a comment by Arnim von Gleich. Edeltraud Günther's and Marco Lehmann-Waffenschmidt's study on "Deceleration – Revealed Preference in Society and Win-Win-Strategy for Sustainable Management" (commented by Fritz Reheis) and Raimund Bleischwitz' paper on "Assessment Criteria for a Sustainability Impact Assessment in Europe" (commented by Kilian Bizer) form the final Part III "Arrangements in Society and Economy Towards Sustainability" of the volume.

On behalf of the authors of this volume I very much acknowledge the financial as well as particularly also the idealistic support by the BMBF and the "RIW front men" Dr. Jens Hemmelskamp and Dr. Alexander Grablowitz who encouraged and inspired us during the whole time. Last, but not least, particular thanks are due to Thomas Krause and Ferri Leberl for formatting the electronic manuscript and to Barbara Feß from the Physica-Verlag for her helpful and friendly support as well as to the Hanse Institute for Advanced Study (Hanse-Wissenschaftskolleg) at Delmenhorst near Bremen for providing comfortable conditions for realizing the last "finish" of the manuscript.

Dresden, March 2007 Marco Lehmann-Waffenschmidt

List of Contributors

Prof. Dr. Frank Beckenbach
University of Kassel
Faculty of Economics
Nora-Platiel-Str. 4
D- 34127 Kassel
beckenbach@wirtschaft.
uni-kassel.de

Prof. Dr. Kilian Bizer
Chair for Economic Policy and SME Research
Economics Department
University of Göttingen
Platz der Göttinger Sieben 3
D-37073 Göttingen
Bizer@wiwi.uni-goettingen.de

Prof. Dr. Raimund Bleischwitz
Co-Director Research Group 'Material Flows and Resource Management'
Wuppertal Institute
PO Box 100480
D-42004 Wuppertal / Germany
Visiting professor at the College of Europe, Bruges/Belgium
Raimund.bleischwitz@
wupperinst.org

Dipl.-Polit. Klaus Burmeister
Z_punkt GmbH
The Foresight Company
Zeche Zollverein
Bullmannaue 11
D-45327 Essen
burmeister@z-punkt.de

PD Dr. Klaus Fichter
Borderstep Institute for Innovation and Sustainability
P.O. Box 37 02 28
D-14132 Berlin
fichter@borderstep.de

Prof. Dr. Arnim von Gleich
University of Bremen
Faculty 4, Production Engineering
Technological Design and Development
Badgasteiner Str. 1
D-28359 Bremen
Gleich@uni-bremen.de

Prof. Dr. Edeltraud Günther
Technical University of Dresden
Department of Economics and Business Administration
Environmental Management
D-01062 Dresden
Bu@mailbox.tu-dresden.de

Dr. Joachim Hafkesbrink
Innowise research & consulting GmbH
Ludgeristrasse 20
D-47059 Duisburg
jh@innowise.de

Dr. Klaus Jacob
Environmental Policy Research Centre
Freie Universität Berlin
Ihnestrasse 22
D-14195 Berlin
jacob@zedat.fu-berlin.de

Dr. René Kemp
United Nations University
Maastricht Economic and social Research and training centre on Innovation and Technology (UNU-MERIT)
Keizer Karelplein 19
NL-6211 TC Maastricht
r.kemp@merit.unimaas.nl

Prof. Dr. Marco Lehmann-Waffenschmidt
Technical University of Dresden
Department of Economics and Business Administration
Managerial Economics
D-01062 Dresden
manaecon@mailbox.tu-dresden.de

Dr. Christian Lutz
Gesellschaft für Wirtschaftliche Strukturforschung mbH
Heinrichstr. 30
D-49080 Osnabrück
Lutz@gws-os.de

Prof. Dr. Bernd Meyer
Fachbereich Wirtschaftswissenschaften
Universität Osnabrück
D-49069 Osnabrück
Gesellschaft für Wirtschaftliche Strukturforschung mbH
Heinrichstr. 30
49080 Osnabrück
bernd.meyer@uni-osnabrueck.de

Dipl.-Vw. Jan Nill
European Commission Joint Research Centre
Institute for Prospective Technological Studies (IPTS) — Unit J03
Support to the European Research Area
Edificio Expo, C/ Inca Garcilaso s/n
E-41092 Sevilla
Jan.Nill@ec.europa.eu

Privatdozent Dr. Niko Paech
Carl von Ossietzky University of Oldenburg
Faculty II
Chair for Strategic and Environmental Management
D-26111 Oldenburg
Niko.paech@uni-oldenburg.de

Prof. Dr. Reinhard Pfriem
Carl von Ossietzky University of Oldenburg
Faculty II
Chair for Strategic and Environmental Management
D-26111 Oldenburg
reinhard.pfriem@uni-oldenburg.de

Dr. phil. habil. Fritz Reheis
Branigleite 19,
D-96472 Rödental b. Coburg
fritz.reheis@t-online.de

Dr. Klaus Rennings
Centre for European Economic Research (ZEW)

Research Area Environmental and Resource Economics,
Environmental Management
mail address: P.O. Box 103 443
D-68034 Mannheim
renninegs@zew.de

Prof. Joachim Schleich, PhD
Fraunhofer Institute for Systems and Innovation Research
Breslauer Str. 48
D-76139 Karlsruhe
Adjunct Professor, Virginia Tech. University,
Blacksburg, VA 24061-0401, USA
j.schleich@isi.fraunhofer.de

Prof. Dr. Stefan Zundel
University of Applied Sciences Lausitz
Department for Informatics, Mechanical and Electrical Engineering and Economics
Großenhainer Str. 57
D-01968 Senftenberg
zundel@fh-lausitz.de

Part I

New Approaches to Environmental Innovation Policy

Windows of Opportunity for Radical Technological Change in Steel Production and the Influence of CO_2 Taxes

Christian Lutz[1], Bernd Meyer[2], Jan Nill[35], and Joachim Schleich[4]

[1] Gesellschaft für Wirtschaftliche Strukturforschung mbH, Heinrichstr. 30, D-49080 Osnabrück.
Lutz@gws-os.de

[2] Gesellschaft für Wirtschaftliche Strukturforschung mbH, Heinrichstr. 30, D-49080 Osnabrück.
bernd.meyer@uni-osnabrueck.de

[3] European Commission Joint Research Centre, Institute for Prospective Technological Studies (IPTS) — Unit J03 Support to the European Research Area, Edificio Expo, C/ Inca Garcilaso s/n, E-41092 Sevilla.
Jan.Nill@ec.europa.eu

[4] Fraunhofer Institute for Systems and Innovation Research, Breslauer Str. 48, D-76139 Karlsruhe, Adjunct Professor, Virginia Tech. University, Blacksburg, VA 24061-0401, USA.
joachim.schleich@isi.fraunhofer.de

[5] The author's contribution results from research carried out at the Institute for Ecological Economy Research (IÖW), Berlin. Views expressed are purely the author's ones.

1 Introduction

The steel industry is one of the most important energy consuming industries. For a given technology of steel production the possibilities of energy saving are rather small, since the used energy carrier is defined and the efficiency of using the produced heat for making steel can hardly be improved. So in this industry the competition between different technologies is at the centre of interest of climate change policy.

The output of the steel industry is produced with different technological concepts, but generally this is done in three stages:

1. Ironmaking: production of pig iron, usually based on the inputs coal, coke, and ore.
2. Crude steel production: purification of iron to steel in oxygen furnaces or melting of scrap steel in electric arc furnaces.
3. Finished steel production: transformation of steel into plates, sheets, tubes, etc.

In Germany, at the moment, two incumbent technologies are competing on the second stage: the basic oxygen furnace (BOF) and the electric arc furnace (EAF). The capital intensive BOF technology produces steel in so-called integrated steel mills with the iron output of the first stage, usually produced from coke and ore in a blast furnace, and gas inputs. The EAF technology can be installed in smaller so called mini-mills and uses electricity and scrap. Since the first step of steel production can be avoided, the production of electric arc furnace steel requires less than half the primary energy demand of the blast furnace-oxygen steel route and CO_2 emissions are much lower. But this is true only for the present time period. At some time in the past scrap must have been produced at the first stage.[6]

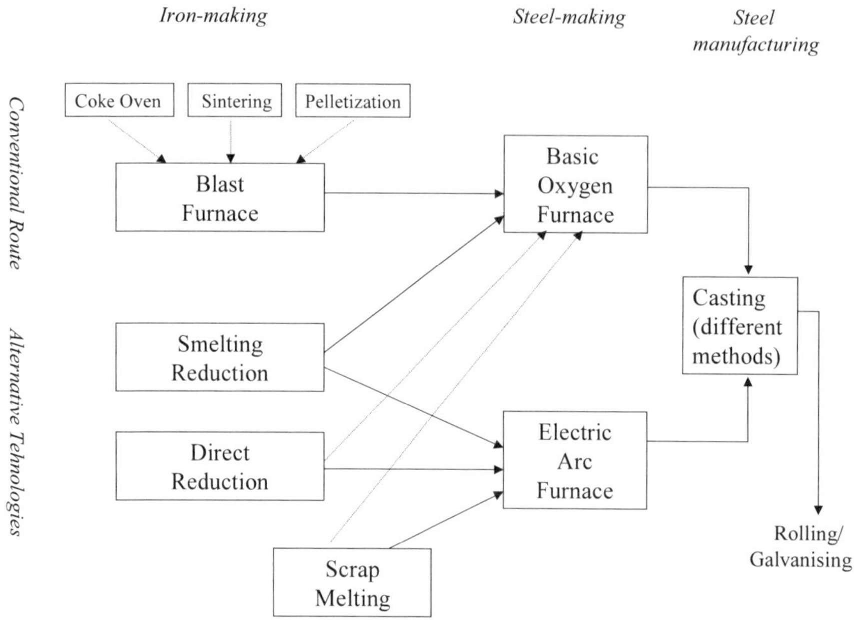

Fig. 1. Schematic lay-out of the iron and steel industry
Source: Luiten (2001, 169), modified

Introducing technology choice into the economic-environmental model PANTA RHEI, [9] and [13] have shown for Germany that the process of technology switch from BOF to EAF, which could be observed in the past, can

[6] To correctly compare energy use of the two processes, the energy embodied in scraps should be accounted for in EAF-steel. To do so, one would have to make assumptions about the number of times steel is recyclable in the various demand sectors. Typically, this type of calculation is not applied when calculating specific energy use of EAF [14].

be strengthened by a CO_2 tax. Likewise, the tax induces technological change towards less energy-intensive processes in both production lines. The present paper asks whether a new technology, which is not yet in the market, might get a better chance of realisation also through climate policy instruments. The smelting reduction technology (SRT) is a new variant of the BOF technology which replaces the traditional first stage, i.e. the coke oven blast furnace (CBF) route of BOF steelmaking, and produces pig iron directly from normal coal and ore in a new type of oven, thus avoiding coke oven operations[7].

According to [7], there have been considerable innovation dynamics of SRT in the Netherlands, even a first commercialisation was intended but has not happened yet. In Germany however, the economic boundary conditions for the introduction of SRT technology appear less favourable since most of the coke ovens and blast furnaces used in the traditional path of the BOF technology can still be used for a long time. Nevertheless, given that substantial reinvestment into coke ovens as important part of the traditional BOF route have taken place in 2003, replacing almost one third of the coke oven capacity, the question arises, whether a window of opportunity for radical technological change has just been closed. In particular, a CO_2 tax may have changed the profitability of the steel production processes in favour of the SRT-BOF technology.

This paper attempts to provide answers using the modified modelling framework PANTA RHEI, which explicitly takes into account the impact of the tax on energy efficiency in the production of BOF and EAF steel. In contrast to the incumbent BOF and EAF technologies, lack of data prevents the direct modelling of the SRT technology with its complete input vector. Instead, we use PANTA RHEI for the calculation of demand and supply effects in the steel sector and the entire economy for the incumbent technologies. These results are then combined via "soft link" with the available information about the SRT technology. Our findings imply that for plausible tax rates there has not been any window of opportunity for the introduction of SRT technology in Germany up to now.

The paper is organised as follows. Section 2 summarises our knowledge of the new SRT technology and the conditions of technological competition. Section 3 gives a short overview of the model PANTA RHEI. In Section 4 we analyse whether climate policy has opened a window of opportunity for SRT in the past or may open such a window in the near future. To do so, we first discuss the reasons for the investment in the old BOF technology in Germany in the past. Then we take a look at the effects of a CO_2 tax on

[7] In principle, SRT could be used to produce hot metal as input into EAF, too. This would allow for the production of higher quality EAF steel, albeit with a higher energy use than scrap-based EAF. For the sake of simplicity and because no reference pilot plant exists until now, this option is excluded in the following analysis and left to future research.

the incumbent technologies for the production of steel in the future. Finally, Section 5 provides conclusions.

2 The Smelting Reduction Technology and the Conditions of Technological Competition in Ironmaking

A direct competitor of the coke oven blast furnace route of BOF steelmaking came up with the so-called smelting reduction technology (SRT). The theory underlying smelting reduction, i.e. converting iron ore directly into crude steel in just one step by using the principle of gasifying coal in a molten bath, has been known since the 1930s. However, notable R&D efforts have only started in 1975. Compared with blast furnaces, basically the sequence of gasification and reduction is changed. Several technology variants which apply this principle have been developed. This ironmaking technology may be combined either with basic oxygen furnaces or with electric arc furnaces.

Smelting technology allows to reduce iron ore to pig iron using coal instead of coke, thus avoiding coke-oven operations. Most SRT devices also omit the agglomeration of iron ore. The process involves both solid-state reduction and smelting, i.e. melting involving chemical reactions. Hence, it comprises two different stages: the pre-reduction unit and the smelting reduction vessel, exploiting the principle that coal can be gasified in a bath of molten iron.

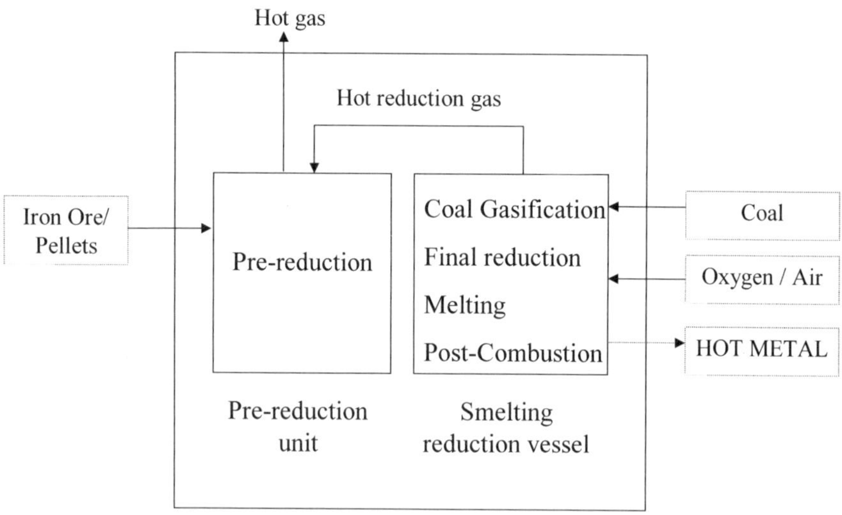

Fig. 2. Schematic lay-out of smelting reduction technology
Source: Luiten (2001, 171)

In the smelting reduction vessel coal is gasified, delivering heat and hot gas containing carbon monoxide, which has a high chemical energy. Heat is used to smelt the iron, whereas the hot gas is transported to the pre-reduction unit to pre-reduce iron-oxides (in a solid state), fed directly into this unit. Subsequently, the pre-reduced iron is transported to the smelting reduction vessel for final reduction. Moreover, carbon monoxide can be oxidised in the smelting reduction in order to deliver additional heat to smelt the iron. This stage of the process is called post-combustion and decreases the reduction potential of the hot gas in the smelting reduction vessel. After post-combustion the hot gas is transported to the pre-reduction unit where the remaining carbon monoxide is used. Since the degree of pre-reduction is determined by the content of carbon monoxide in the hot gas, there is a trade-off between pre-reduction and post-combustion [7, p. 172].

High levels of pre-reduction are characteristic of the first generation processes. The first commercial application and best known example of these is the COREX process. High levels of post-combustion determine second generation processes, the lower degree of pre-reduction that need less coal, because extra heat is generated and used in pre-reduction. However, SRT is not a homogenous technology; there is a variety of smelting reduction processes, whereas only one operates on a commercial basis [7, p. 173]. One of the most promising in terms of energy efficiency and CO_2 emission reduction is the Dutch CCF SRT which was developed in the late 1980s and early 1990s. A small pilot plant was installed and a large demonstration plant was envisaged to be built in the Netherlands. However, it has not been realised yet (for further details, see [7, 11]). As for emission, it is estimated that the CO_2 emissions from this SRT will be 15% lower than the CO_2 emissions of the coke oven blast fournace route [16].

In particular, if technologies are characterised by important increasing returns to adoption such as scale, learning, and network effects, however, technological substitution in favour of cleaner technologies is not an easy or automatic process. Zundel et al. [17, 18] argue that in stable phases in which a certain technology dominates and exploits increasing returns, usually only incremental changes take place. Only in instable phases of technological competition there is a techno-economic "window of opportunity" for competing radical innovations.

According to [11] iron and steel production is a case in point. The coke-oven blast furnace route of BOF steelmaking still dominates the higher quality segment of steelmaking and has exploited scale and learning effects in an impressive way. It is characterised by huge installations and a tremendous capital intensity. Hence, besides technological breakthroughs and development progress of new technologies sunk costs incurred by the dominant technology are an important determinant of windows of opportunities for technological competition. An important indicator for the dynamics of sunk costs are investment cycles. In the Netherlands and in Japan, such a techno-economic window was anticipated due to rapid progress of the new SRT and substantial reinvestment needs for coke ovens.

In Germany, however, such reinvestment needs have been less important. In the year 2003 there have been 15 blast furnaces in Germany with a capacity of 31 Mio. tons. Equipment with a capacity of 8 Mio. t is younger than 10 years (installed in 1993–2003), equipment with a capacity of 7 Mio. t is between 10 and 30 years old, equipment with a capacity of 11 Mio. t is between 30 and 35 years old, and equipment with a capacity of 5 Mio. t is older than 35 years (installed in 1950–1966). In the year 2003 there have been five coke ovens in Germany to produce the inputs for the blast furnaces, the usual lifetime being about 40 years. Four of them have been built between 1984 and 1985 and one with about 30% of the whole capacity has started production in the year 2003[8].

Since blast furnaces can be used longer than 60 years there seems to be no driver for a window of opportunity for this part of the conventional production route. In the case of the coke ovens we have now an even clearer situation. Still, the question arises whether some five or six years ago, when the new coke oven was planned and SRT was at the edge of commercialisation, there might have been a window of opportunity that could have been opened by environmental policy.

According to [3, p. 133] average energy consumption per ton of steel in a conventional integrated steel mill with coke oven blast furnace BOF technology amounts to 19 GJ. For a comparison, however, the reference should be the most efficient incumbent process. On this basis, a comparison of specific energy consumption for both the coke oven blast furnace and the smelting reduction route of BOF processes is provided in Table 1.

The data for the coke oven blast furnace variant of BOF technology is based on very efficient integrated steel mills from Hoogovens in the Netherlands while the data for the SRT variant of BOF technology is based on design studies and a pilot plant. For a comparison of the coal input it has to be considered that SRT only needs steam coal as input which is about 10% cheaper than metallurgical coal [3, p. 160].

For the calculation of specific energy use it is assumed that, in a combined cycle plant, gas and steam are transformed into electricity which is used inhouse for the production of oxygen. The rest is given into the electricity grid. Here, two scenarios are discussed which both assume an efficiency of the steam transformation of 35%:

- Scenario low: lower heating value for coal of 29 GJ/t; electric efficiency of gas transformation 60%; efficiency of electricity production 60%
- Scenario high: higher heating value for coal of 32 GJ/t; electric efficiency of gas transformation 45%; efficiency of electricity production 40%

[8] Source: VDEh databank PLANTFACTS, November 2003

Table 1. Comparison of BOF/CBF and BOF/SRT technologies

	BOF (coke ov. blast furn.)			BOF (SRT)		
	Energy inputs	GJ/thm low	GJ/thm high	Energy inputs	GJ/thm low	GJ/thm high
Coal	0,59 t	17,1	18,9	0,64 t	18,6	20,5
Other fuels		2,1	2,1			–
Oxygen		–	–	0,67 t	1,1	1,7
Electricity	69 kWh	0,4	0,6		–	–
Export gas	-3,7 GJ	-3,7	-4,2	-3 GJ	-3,0	-3,4
Export heat		–	–	-5,78 GJ	-3,4	-5,1
Specific energy input in GJ/thm		15,9	17,4		13,3	13,7
Investment	US$/thm	385		US$/thm	150–180	
Variable cost		84,6–109,5			70–90	
Total cost		121,7–160,1			90–115	

Source: De Beer et al. (1998)

For the economic figures, assumptions about the calculation of annuities of investment are important: the investment is depreciated over 15 years, the real interest rate is taken alternatively as 5% (lower bound) and 10% (higher bound). The investment in conventional technology is split into 195 US$/thm for the blast furnace, 145 US$/thm for the coke oven and 45 US$/thm for the sintering equipment.

We are not able to comment in detail on the calculations of de Beer et al. (1998)[3], but Table 1 clearly shows that total cost per t of hot metal is, in the SRT case, only about 75% (90/121.7 or 115/160.1) of the costs of the conventional technology. But this is not the comparison to be made because the investment decision addressed concerns the coke oven only, which represents 37,7% of total investment[9]. So in his comparison the investor will calculate with reduced total costs of 98.5 to 128.5 US$ in the conventional case against 90 to 115 US$ in the SRT case. The advantage of the SRT technology shrinks to 10% which may not be enough to cover the risks of a switch to a totally new technology. Apparently, this has been the case in the concrete investment decision we are exploring. Furthermore, this small discrepancy may not be significant since the data are based on estimates only – although the cost data reported for the newly installed German coke oven is similar[10].

So at the time of the investment decision, which may have taken place in 1997 or 1998, it seems to have been economically rational to invest in the incumbent technology. However, if at the end of the 1990s there had been a tax

[9] For an exact comparison, the costs of retrofitting of blast furnaces, which takes place regularly, need to be integrated, however, quantitative data are missing.

[10] The actual coke oven investment amounts to 800 Mio. Euro for an annual capacity of 2.5 million tons of coke (MaschinenMarkt, April 28th, 2003).

on CO_2 in place, it would have made sense to invest in SRT. For a profitability assessment, information about the impact of the carbon tax per ton of hot metal for the conventional and the new SRT BOF as well as the incumbent EAF technologies is required. For the conventional variant of BOF and EAF technology we can calculate these figures using the economic-environmental model PANTA RHEI which also takes into account the impact of the tax on specific energy use in the incumbent processes. For the SRT variant of the BOF technology we can estimate these figures roughly by comparing the different energy inputs of both technologies. Furthermore, the advantage of SRT concerning energy inputs – the production of heat – has to be taken into account.

3 The Model PANTA RHEI

PANTA RHEI – the name means "all things flow" and stems from the Greek philosopher Heraklit – is an environmentally extended version of the econometric simulation and forecasting model INFORGE (INterindustry FORecasting GErmany). Its performance is founded on the INFORUM philosophy [1], which maintains that econometric input-output models should be constructed in a bottom-up and fully integrated manner. Here "bottom-up" means that each sector of the economy has to be modelled in great detail and that the macroeconomic aggregates have to be calculated by explicit aggregation within the model. The construction principle "fully integrated" means that the model structure takes into account a variable input-output structure, the complexity and simultaneity of income creation and distribution in the different sectors, its redistribution among the sectors and its use for the different goods and services which the sectors produce in the context of global markets.

INFORGE consistently describes the annual inter-industry flows between 59 sectors, their contributions to personal consumption, government, equipment investment, construction, inventory investment, exports as well as prices, wages, output, imports, employment, labour compensation, profits, taxes, etc. for each sector and for the macro economy.

The economic part of the model also contains a complete SNA system to calculate the aggregated variables and the income redistribution between the government, households, firms, and the rest of the world. For these institutional sectors, disposable income and flow of funds can be estimated and the budget of the government, including fiscal policy and the social security system, is depicted endogenously. In this way, the model provides a consistent framework for the analysis of market-based climate change policies, as indirect effects an other industries are captured and additional tax revenues are adequately accounted for.

In addition, PANTA RHEI contains a deeply disaggregated energy and air pollution module which distinguishes 30 energy carriers and their inputs in

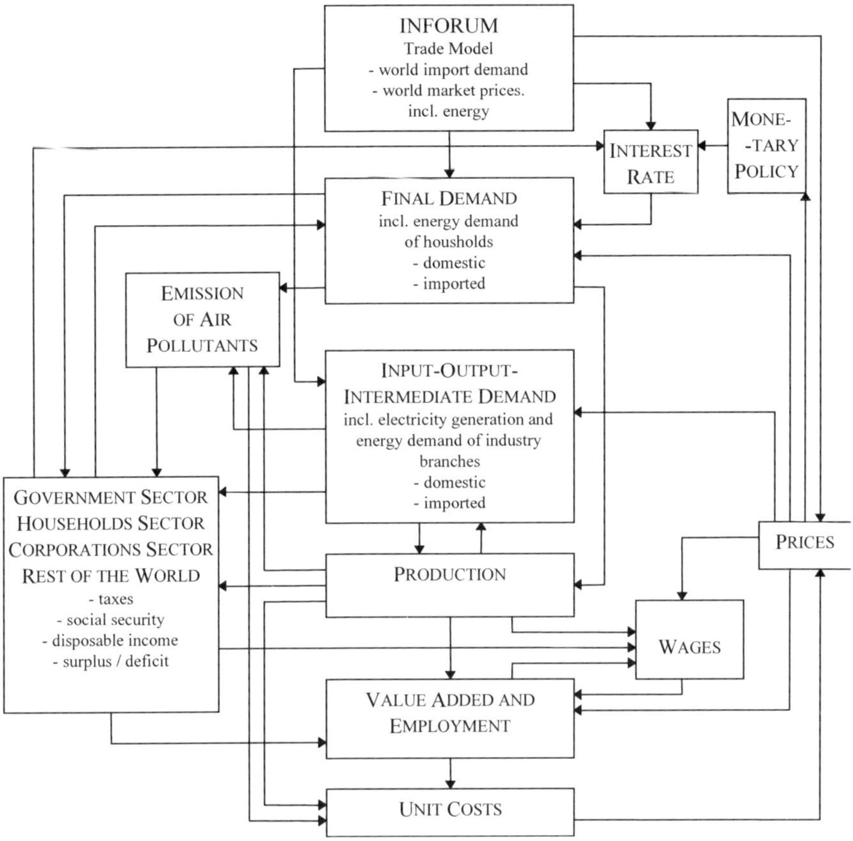

Fig. 3. The model structure of PANTA RHEI

121 production sectors and households as well as the related CO_2 emissions. Energy demand is fully integrated into the intermediate demand of the firms and the consumption demand of households. Energetic input coefficients are generally explained by relative prices and trends.

The supply of nuclear energy and renewable energy for electricity production is modelled exogenously since they primarily depend on policy decisions in Germany. As for the transport sector, the gasoline and diesel demand of households and firms are calculated using an extended road traffic module which explains the stock of cars and trucks and their usage as well as technical progress in the new vehicle vintages.

Parameters in all equations in PANTA RHEI are estimated econometrically using OLS.[11] The model has been used in many studies to explain struc-

[11] Of course, from a theoretical point of view, simultaneous equation estimation techniques would have to be applied. However, due to the large number of about

tural effects of environmental policy measures, to forecast energy and carbon emissions, and to explain the effects of abatement techniques on emissions and the economy [2, 10, 8].

In the conventional PANTA RHEI model technical change is not directly depicted. Rather the result of this process – changing input structures – is shown in time series of input-output tables. This allows for a reduced-form type estimation of price-dependent input coefficients, but there is no link to the underlying technologies.

In this paper we follow [13] and [9] and no longer regard technological change and changes in production processes, which translate into changes in the input-output coefficient, as a black box. For the steel sector, we chose a much more disaggregated structural-form type approach and explicitly model the main production processes BOF and EAF, the choice between these production processes, and the development of energy intensity for the respective best-practice technologies. This also allows for a more realistic analysis of policy scenarios where the effects can be traced down to individual processes.

The econometric input-output model PANTA RHEI implies – in contrast to general equilibrium models based on CES functions – limitationality of the input factors in the individual branches. The input coefficients are modelled as price-dependent which is then interpreted, not as the result of substitution, but of cost-induced technological progress, which results in changes in the choice of process. In the new modelling approach, this is linked to actual production processes to allow for an integrated bottom-up/top-down analysis. To do so, among others, investments, production amounts, detailed input structures and the process-specific input demand of the respective best-practice technologies (*trajectories*) are determined for the historical observation period 1980–2000 for the different process lines (*paradigms*) [4, 5]. Based on these data, the paradigm-specific investments, i.e. the choice of technology and the development of technical change in the model can be estimated econometrically as a function of prices and other variables. The revealed correlations serve as the basis for the ex-ante policy simulations.

4 Could a Carbon Tax Open a Window of Opportunity?

How could the introduction of a carbon tax have influenced the investment decision between the conventional coke oven route and the SRT process? We will try to answer this question in two steps: In a simple static calculation for the year 1997 we first compare the possible direct impact of a CO_2 tax with the overall investment costs of the new coke oven. We then use model simulations to support our findings and to make a dynamic assessment.

When we follow the argument of [16] that CO_2 emissions of SRT are 15% lower in the Dutch case than in the conventional BOF route, annual cost for

5000 estimated variables in PANTA RHEI this is not feasible. Model specification is based on conventional hypothesis testing (t-statistics, R^2).

CO_2 emissions would have been 3 Euro/thm lower for SRT – assuming quite high tax rates of 20 Euro/t CO_2. If the investor had included these additional costs for a long period of 15 or even 20 years with a discount rate of 5%, the additional cost for CO_2 tax of about 35 to 50 Euro/thm would have been substantial in comparison to investment costs of about 100 to 130 Euro/thm. If the assumption of Worrell et al. (1997)[16] had also been valid for Germany at the end of the 1990es, a CO_2 tax indeed might have opened a window of opportunity.

But how about the German case at that time? The model PANTA RHEI includes data of the German Federal Statistical Office for this period. Table 2 shows for the year 1997 the needed coal input in PANTA RHEI, which is the average over all German conventional BOF production, and the assumptions of [3] for the best available BOF blast furnace and SRT technologies in 1997. The figures for conventional BOF fit quite well, as the authors of [3] describe the best-practice technology, and PANTA RHEI contains the actual German average.

Table 2. Ratio of coal input in t per t of hot metal (thm) in 1997

	PANTA RHEI	de Beer	de Beer
	BOF (CBF)	BOF (CBF)	BOF (SRT)
steel production	0,51		
coke oven (33% ass.)	0,17		
sum	0,67	0,59	0,64

If we further assume that the investment decision was based on these relations of coal input to hot metal output, a tax of 20 Euro/t CO_2 would have changed annual energy costs of the plant under discussion in the following way: On the one hand, annual additional costs for coal input of SRT in comparison to conventional BOF would have amounted to about 3 Euro/thm. On the other hand, the costs for electricity needed for the coke oven increase by about 1 Euro/thm. SRT could have produced additional heat for about 2.5 Euro/thm. So, the overall advantage in Germany has only been about 0.5 Euro/thm. An important reason for the difference to the assumptions of [16] are the structure and level of German electricity and heat production and energy prices. Since, at that time, about 40% of electricity were produced carbon-free from nuclear and, to a much lesser extent, from renewable energy sources. And since end-user prices were much higher than industry prices, the steel industry would probably have expected – according to econometric estimations in PANTA RHEI – electricity and heat prices to react much less to the tax than prices of subsidized coal (a finding in many simulations about CO_2 taxes in Germany). This also implies, that the end-users, who apparently tend to exhibit a relatively low price elasticity, would have paid the biggest part of the CO_2 tax. Therefore, such a tax being in place at the end of the 1990ies would probably not have opened a window of opportunity for a technology with a higher coal input, even if the latter is overcompensated by energy savings in other domains. In contrast, a combined CO_2 and energy tax, as planned by the EU commission at the beginning of the 1990ies, would have favoured SRT more.

To evaluate the dynamic and long-term impacts of price instruments, results of two simulations with PANTA RHEI – a base scenario and a tax policy scenario – are used [9]. In the policy scenario, a CO_2 tax – as part of a global CO_2 tax or emissions trading system – is introduced in progressive stages starting in 2005. The CO_2 tax is introduced in 2005 and increases from 5€to 20€per ton CO_2 in 2010. This is equivalent to a price per ton of carbon of more than 73 Euro in 2010. Thus, the CO_2 tax lies in the range of recent model estimates for CO_2 market prices [15]. The CO_2 tax is levied on all fossil energy carriers according to their carbon content, so that the use of coal is more heavily taxed than oil or gas.

Table 3. Ratio of coal input in t per t of steel output in 2010

	PANTA RHEI	de Beer
	BOF (CBF)	BOF (SRT)
steel production	0,43	
coke oven (33% ass.)	0,14	
sum	0,57	0,64

Both simulations also reveal important information for the conventional route of BOF steel production. The base simulation shows technical progress for the BOF(CBF) technology until 2010 due to induced R&D in the supplier sectors. Direct coke and coal input will decrease from 0.51 t per t of hot metal output in 1997 to 0.43 t in 2010. Assuming a ratio of additional coal use in the coke oven process of 33% – given in PANTA RHEI – overall average coal input ratio per ton of hot metal will be 0.57 in 2010 in contrast to 0.67 in 1997. According to [3] the expected ratio is 0.64 for the SRT process.

Table 4. CO_2 emissions in t per steel output in t

	Base	Tax	de Beer	Tax
	BOF (CBF)	BOF (CBF)	BOF (SRT)	EAF
1997	2,09	2,09	1,99	1,18
2010	1,78	1,69	1,99	0,74

What can we expect about technical progress of the potential SRT technology? Lutz et al. [9] argue that the assumption of Pavitt (1984) of "supplier-dominated firms" is suitable for the German iron and steel industry. The improvement of best-practice technologies is mainly driven by input price relations and R&D spending of supplying industries according to the investment demand of steel producers. Without investment in SRT in Germany or other countries in the period until 2010, we cannot expect much technical progress. So, if it is assumed that a comparable technical progress for a single new SRT process is unlikely or takes place more slowly than for CBF, CBF will be less coal intensive than SRT already in the base scenario in 2010 (see Table 3). Thus, while our analyses do not reveal whether there was a window of opportunity for SRT in 1997, our findings suggest that in 2010 it will definitely be closed.

Even though the detailed numbers of the different BOF technologies have to be regarded with some caution, the simulation results also show clearly that a CO_2 tax will neither favour BOF/CBF nor SRT in the long run: a CO_2 tax will in the first place induce future investment into EAF due to large differences in the CO_2 intensity. Table 4 shows the differences in CO_2 emissions per t of steel output in 2010, which will still be high if we consider additional emissions of EAF, embodied in the scrap. But the shift from BOF to EAF takes a long time. It will take until 2020 to shift capacities in a magnitude we have disussed above. In fact, in 1997 this argument did not hold. One important reason is that BOF and EAF steel are only partly substitutes, as the quality of BOF steel is higher. The sudden replacement of 1/3 of the German BOF capacity in 2003 would have been impossible.

5 Conclusions

We briefly summarise the main results of our paper:

1) The results generally highlight the importance of taking into account the effects of policy-induced technological change on incumbent technologies when exploring windows of opportunities for new technologies which may be opened by policy interventions. Indeed, as already noticed by [11], windows of opportunity are a phenomenon of technological competition and the progress of the incumbent technology has to be taken into account. It is sometimes even fuelled by revived competition, known in the literature as "sailing ship effect". The progress of the conventional coke oven route in the 1990s was one reason that some conventional steel producers did stop investment into R&D of new technologies.
2) More specifically, our findings suggest that a CO_2 tax of the magnitude assumed might not be sufficient to render future investments in the SRT technology profitable compared to incumbent steel producing processes
a) because differences in CO_2 emissions compared to BOF steel produced via the coke oven blast furnace route are not sufficiently large in the German case and even seem to become smaller because of induced improvements in energy efficiency in incumbent processes; and

b) because the tax essentially favours EAF steel more than any variant of BOF steel in the long run. Hence, the window of opportunity for environmentally beneficial technological competition in BOF steelmaking in integrated steel mills may have vanished for quite a long time.

3) In the future the more interesting question will be whether a CO_2 tax or emission trading will enhance another kind of technological competition. Will such a tax induce the challenge of the conventional BOF route by an upgraded EAF technology with hot metal input, e.g. by SRT? To answer this question on solid grounds, a considerable modelling effort to integrate all environmental and economic aspects of this potential competition would be necessary which is beyond the scope of this paper. Whether the EU CO_2 emissions trading system, which is scheduled to start in 2005 for energy intensive companies, spurs such a competition is in principle an open question and depends on the specifics of the national allocation plans. In countries where incumbent companies, like existing BOF steel producers, receive all allowances in the primary allocation for free (grandfathering) and new entrants, like new EAF/SRT steel producers, have to buy their allowances on the market or through an auction, such a competition would be difficult to arouse [11]. However, the emission trading system may favour competition too: new entrants may also receive allowances for free, plant operators may transfer allowances from closures to new installations, or incumbents may keep allowances from closures for the future [6].

References

1. Almon C. (1991): The INFORUM Approach to Interindustry Modeling. Economic Systems Research 3 (1), 1–7
2. Bach S., Kohlhaas M., Meyer B., Praetorius B. and Welsch H. (2002): The effects of environmental fiscal reform in Germany: a simulation study. Energy Policy 30 (9), 803–811
3. De Beer J., Blok K. and Worrell E. (1998): Future Technologies for energy-efficient iron and steel making. Annual Review of Energy and Environment 23, 123–205
4. Dosi G. (1982): Technological paradigms and technological trajectories: A suggested Interpretation of the determinants and directions of technical change. Research Policy 11, 147–162
5. Dosi G. (1988): The nature of the innovative process. In: Dosi G., Freeman C., Nelson R., Silverberg G. and Soete L. (eds.): Technical Change and Economic Theory. Pinter Publishers, London, New York, 221–238
6. Graichen P. and Requate T. (2005): Der steinige Weg von der Theorie in die Praxis der Emissionshandels: Die EU-Richtlinie zum CO_2-Emissionshandel und ihre nationale Umsetzung. Perspektiven der Wirtschaftspolitik 6 (1), 41–56
7. Luiten E. (2001): Beyond Energy Efficiency. Actors, networks and government intervention in the development of industrial process technologies. Universiteit Utrecht, Utrecht

8. Lutz C. (2000): NO_x Emissions and the Use of Advanced Pollution Abatement Techniques in West Germany. Economic Systems Research 12 (3), 305–318
9. Lutz C., Meyer B., Nathani C. and Schleich J. (2005): Endogenous technological change and emissions: the case of the German steel industry. Energy Policy 33, 1143–1154
10. Meyer B. (2001): CO_2-Taxes, Growth, Labor Market Effects, and Structural Change – An Empirical Analysis. In: Welfens P.J.J. (ed): Internationalization of the Economy and Environmental Policy Options. Springer, Berlin, 331–352
11. Nill J. (2005): Technological competition, time, and time windows: the case of iron and steel production technologies. In: Sartorius C. and Zundel S. (eds.): Time Strategies, Innovation, and Environmental Policy. Cheltenham, Edward Elgar, 255–286
12. Pavitt K. (1984): Sectoral patterns of technical change: Towards a taxonomy and a theory. Research Policy 13, 343–373
13. Schleich J., Nathani C., Ostertag K., Schön M., Walz R., Meyer B., Lutz C., Distelkamp M., Hohmann F. and Wolter M.I. (2002): Innovationen und Luftschadstoffemissionen – Eine gesamtwirtschaftliche Abschätzung des Einflusses unterschiedlicher Rahmenbedingungen bei expliziter Modellierung der Technologiewahl im Industriesektor. Dokumentation Stahlindustrie, ISI/GWS, Karlsruhe, Osnabrück, April 2002
14. Schön M. and Ball M. (2003): Eisen und Stahl. Sector report for "Werkstoffeffizienz – Systemanalyse zu den Kreislaufpotenzialen energieintensiver Werkstoffe und ihrem Beitrag zur rationellen Energienutzung". Final Report for the Federal Ministry of Economics and Labour (BMWA = Bundesministerium für Wirtschaft und Arbeit), Fraunhofer ISI, Karlsruhe
15. Springer U. and Varilek M. (2004): Estimating the price of tradable permits for greenhouse gas emissions in 2008–12. Energy Policy 32, 611–621
16. Worrell E., Bode J.-W. and de Beer J. (1997): Energy Efficient Technologies in Industry, the ATLAS Project. Department of Science, Technology & Society, Utrecht University, Report No. 97001
17. Zundel S., Erdmann G., Nill J., Sartorius C. und Weiner D. (2003): Innovation, Zeit und Nachhaltigkeit – Zeitstrategien ökologischer Innovationspolitik – der Forschungsansatz. In: Horbach J., Huber J. und Schulz T. (eds.): Nachhaltigkeit und Innovation. Rahmenbedingungen für Umweltinnovationen, ökom Verlag, München, 55–88
18. Zundel S., Erdmann G., Kemp R., Nill J. and Sartorius C. (2005): Conceptual Framework. In: Sartorius C. and Zundel S. (eds.): Time Strategies, Innovation, and Environmental Policy. Cheltenham, Edward Elgar, 10–54

Comment: Approaches to the Modelling of Innovations for Sustainable Economic Systems

Klaus Rennings

Centre for European Economic Research (ZEW), Research Area Environmental and Resource Economics, Environmental Management, P.O. Box 103 443, D-68034 Mannheim.
rennings@zew.de

1 Survey of Various Model Types

Until recently political and economic approaches of environmental innovation have primarily drawn on factors such as "market pull", "technology push", or "regulatory push/pull" [6]. Endogenous potentials and company-specific determinants have played a subordinate role in studies to date.

The different modelling approaches and applications can be assigned to a number of different levels of aggregation. Case study approaches on the one hand (refer for example to [2]) draw on companies and value chains. The corporate and value chain related approaches (micro/meso level) adopted to date have been exploratory in nature and largely restricted to case analyses. In this context, relevant external and internal determinants can be identified, interrelations specified and "if-then hypotheses" generated. This background emphases the value of work undertaken to date, even though the paucity of case examples prevents empirically founded generalisations and invalidates their universal applicability for entire groups of companies or industries.

Econometric approaches (refer for example to [7]) on the other hand deal with entire groups of companies (e.g. firms operating environmental management systems) for which generally valid findings are sought. This means that theoretical hypotheses, or hypotheses based on the findings of case studies, can be assessed with the help of innovation-oriented econometric models. The advantage of this approach is that factors which are specifically influenced by companies themselves (e.g. the role of market strategies or individually designed corporate environmental management systems) can be included alongside external factors (such as market structure and position in the value chain). There is a particular need for research on the collection of representative samples (control group) and the study of mutual causalities (e.g. how environmental management influences innovation and vice versa). The principal limits of micro-econometric approaches relate to the determination of indirect effects on the macroeconomic level (typically price and substitution effects).

In the case of evolutionary approaches (refer for example to [9]) the object of study is usually a line of technological development such as fuel cell technology. Evolutionary approaches offer a broad and open theoretical framework for innovation processes and are particularly useful when it comes to mapping the properties of innovation processes (e.g. non-deterministic processes, path dependence, uncertainty, situation and context relatedness, coincidences) which are neglected in economic approaches based on a more restricted set of assumptions. The more quantitative an evolutionary approach is, the better it is for the purpose of deriving policy-relevant conclusions. Research also needs to be undertaken through closer links between evolutionary and neoclassical approaches (e.g. integration of learning by doing or path dependencies in neoclassical models).

Econometric input-output models, such as that of Lutz et al. (see page 3–17) and computable general equilibrium models (CGE models; c.f. the survey by [3]) are positioned at a further level of aggregation. Whether technical progress can be explained by exogenous or endogenous factors plays a central role in both model types.

Econometric input-output models can be used to map the influence of policy measures (e.g. the introduction of a CO_2 tax) on specific industries and the economy as a whole, and to record this influence in terms of the relevant sustainability indicators (e.g. CO_2 emissions). As a result, these models can be used for policy decisions in order to increase the reliability of intended measures (i.e. to calculate if intended targets can be reached). Alongside direct effects, these models also record indirect effects on the economy. One problem, however, relates to the availability of valid data (time series, etc.) and the validity or transferability of behavioural assumptions (e.g. corporate investment decisions) between "manageable" decision situations (e.g. the steel industry) and highly complex and dynamic decision making contexts, such as in young, turbulent fields of technology.

Among macroeconomic approaches, computable general equilibrium models have become standard instruments for estimating the overall economic impact of policy measures in the field of climate protection (reduction of greenhouse gases). A broad array of different approaches to the modelling of technical progress is now available. The endogenous treatment of technical progress has to date been highly dependent on ad-hoc assumptions, however. These models also need to be developed to enable them to take account of path dependencies, the uncertainties inherent in new technologies, the heterogeneous nature of business behaviour, and investment incentives.

More work needs to be done at every level of aggregation on all models in terms of their information content and relevance for action by policymakers and other players: clear target, or sustainability indicators are required as reference values on the output side in each case.

It is important that the three aggregation levels touched on and the associated model approaches are all legitimate in themselves and need not necessarily be systematically and artificially linked together. Attempts to achieve structural harmony between all three levels, or to link them smoothly and fully together could result in the models forfeiting their individual creative virtues. Nonetheless, certain aspects of each of the approaches could be fruitfully brought together and, in fact, there is scope for mutual learning in certain selected areas. Micro approaches could, for example, help to map the behaviour of companies more precisely and realistically (e.g. by classifying types of investment decisions or ways of dealing with the risks inherent in new technologies) and, by means of disaggregation, enhance the explanatory and forecasting power of the models.

On the other hand, econometric and macroeconomic models can also fruitfully complement case studies at the micro and meso levels by evaluating the hypotheses generated at these levels in relation to a statistically significant number of companies and by thus substantially extending the validity of the conclusions reached. One such example is the study of the innovation impact of environmental management systems previously referred to. Case studies and econometric approaches have already been successfully linked in this area. In order to map indirect, i.e. price, substitution and income, effects a further link would need to be made to macroeconomic models. Figure 1 shows the various model levels and potential links between them.

2 The Paper by Lutz et al. on Windows of Opportunity for Radical Innovations in Steel Production and the Influence of CO_2 Taxes

Lutz et al. [4] focus on the issue of windows of opportunity and the timing of technology developments and decisions. The window of opportunity concept is derived from the evolutionary innovation economics of [1] and [5]. This concept is based on the observation that, owing to lock-in effects, the diffusion of new technologies, particularly in the field of major technologies and the framework of technological trajectories, is scarcely possible.

A window of opportunity only opens if the existing technology development system becomes unstable and thus allows for new technologies to hit the market. Windows of opportunity may, for example, open in the following situations:

- Reinvestment cycles, i.e. replacement and expansion requirements with regard to the dominant technology
- Politically motivated stipulations or laws (e.g. Large Combustion Plant Directive) which bring about changes in the usual renovation cycle.

The paper by Lutz et al. [4] examines whether there was a window of opportunity for radically new steel production technology (SRT, or smelting reduction

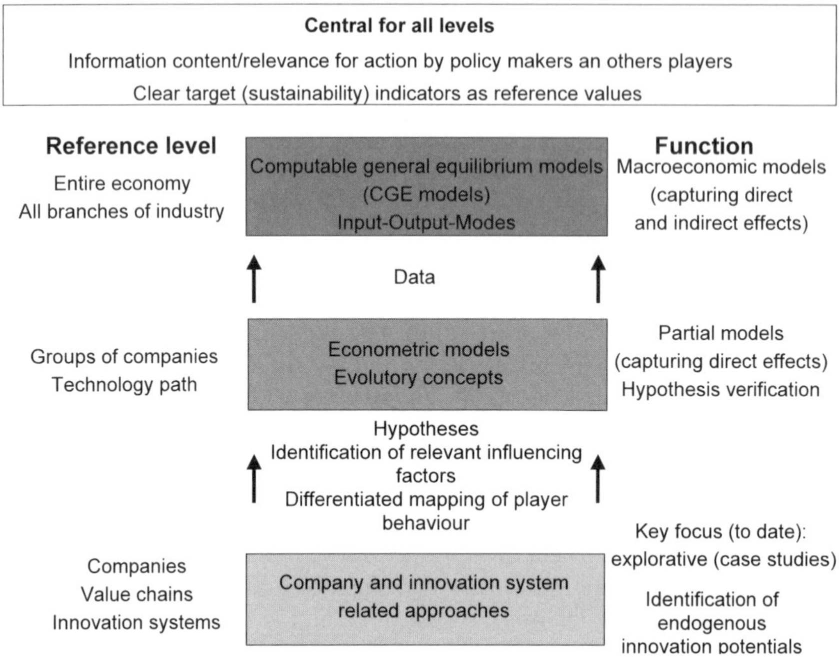

Fig. 1. Various levels of model concepts of sustainable innovation and potential links between them

Source: Rennings K. and Fichter K. (2003) [8].

technology) surpassing the traditional technology (BOF, or basic oxygen furnace) at the end of the 1990s, given the high investment requirements at this time and bearing in mind that a CO_2 tax might well have supported a breakthrough by this new technology.

The authors come to the conclusion that there was no window of opportunity for the new technology, at least not in Germany. There are three key reasons for this conclusion:

- While SRT technology reduces steel making costs by 25% with regard to blast furnaces, this process only accounts for around one third of total required investment. Only 10% of total investment costs are therefore saved, and this represents a relatively weak incentive for a radical technological change.
- In contrast to other countries (the Netherlands and Japan are referred to in the paper) in which a window of opportunity did indeed open for the new technology, around 40% of electricity is generated in Germany by nuclear power stations with no CO_2 output. The new technology would have increased coal and reduced electricity input so that a CO_2 tax in Germany would not have represented a decisive advantage.

- Furthermore, it may also be assumed that the traditional technology will make efficiency progress regarding the amount of coal input per ton of produced steel. This will erode the environmental advantages of the new technology entirely.

The case analysed in the paper is a good illustrative example of the technological aspects of innovation decisions. Outcomes and recommendations will differ depending on whether and how the factors described above (required input of coal or electricity, technical progress of traditional technologies) are taken into account in the model.

One deficit of the paper, however, is that the example seems very contrived dealing as it does with the hypothetical issues as to whether a window of opportunity was, or was not open which has since been closed by replacement investments. The discussion of a third technology (EAF, or electric arc furnace) based on the use of scrap is also confusing. This technology is only of limited value in terms of replacing BOF given that the use of scrap is always dependent on the production of a certain amount of primary steel. In this respect, the need for research identified by the authors must be underlined, for instance should the diffusion of the three technological alternatives BOF vs. SRT vs. EAF be analysed in a single integrated simulation.

References

1. Dosi G. (1982): Technological Paradigms and Technological Trajectories: a suggested Interpretation of the Determinants of Technical Change. Research Policy, Vol. 11, 147–162
2. Klemmer P., Lehr U. und Löbbe K. (1999): Umweltinnovationen: Anreize und Hemmnisse. Berlin, Analytica Verlag
3. Löschel A. (2002): Technological Change in Economic Models of Environmental Policy: A Survey. Ecological economics, 43, 105–126
4. Lutz C., Meyer B., Nill J. and Schleich J. (2005): Windows of Opportunity for Radical Technological Change in Steel Production and the Influence of CO_2-Taxes. In this volume
5. Nelson R. and Winter S. (1982): An Evolutionary Theory of Economic Change. Cambridge
6. Rennings K. (2000): Redefining Innovation – Eco-Innovation Research and the Contribution from Ecological Economics. In: Ecological Economics 32: p. 319–332
7. Rennings K., Ziegler A., Ankele K., Hoffmann E. and Nill J. (2003): The Influence of the EU Environmental Management and Auditing Scheme on Environmental Innovations and Competitiveness in Germany: An Analysis on the Basis of Case Studies and a Large-Scale Survey. ZEW Discussion Paper No. 03-14, Mannheim
8. Rennings K. und Fichter K. (2003): Ergebnisprotokoll des INA-RIW-Workshops "Innovationsmodelle als Grundlage zur Erklärung der Entstehung, Durchsetzung und Wirkung von Nachhaltigkeitsinnovationen" July 11, 2003 at the Centre for European Economic Research (ZEW), Mannheim, Mimeo

9. Zundel S., Erdmann G., Nill J., Sartorius C. und Weiner D. (2003): Zeitstrategien ökologischer Innovationspolitik – der Forschungsansatz. In: Horbach J., Huber J. und Schulz T. (Hrsg.): Nachhaltigkeit und Innovation – Rahmenbedingungen für Umweltinnovationen. Ökom Verlag München, 55–88

Environmental Innovation Policy. Is Steering Innovation Processes Possible?

René Kemp[1] and Stefan Zundel[2]

[1] United Nations University, Maastricht Economic and social Research and training centre on Innovation and Technology (UNU-MERIT), Keizer Karelplein 19, NL-6211 TC Maastricht.
r.kemp@merit.unimaas.nl

[2] University of Applied Sciences Lausitz, Department for Informatics, Mechanical and Electrical Engineering and Economics, Großenhainer Str. 57, D-01968 Senftenberg.
zundel@fh-lausitz.de

1 Introduction

For evolutionary economics, environmental innovation policy involves a serious problem: how to support something that is essentially new and cannot be predicted in advance such as innovations. Many evolutionary economists see the need for a policy improving the scientific and technological infrastructure. Like other economists, however, they are sceptical about possibilities for the government to coordinate actual innovation change processes. Economic development is seen as a basically non-controllable, open-ended process. F. A. Hayek who was likely the leading supporter of this position, held that policy should refrain from any usurpation of knowledge that is not possible to receive in an open society [18, p. 225]. Hayek himself admitted that some kind of prediction pattern might be possible, but in his work this notion is not very well developed, since he and many of his successors in the tradition of the Austrian evolutionary economics rather emphasized the open nature of complex systems such as markets.

Due to this knowledge restriction, it seems impossible to create a more environmentally benign allocation by political means. In particular, it appears impossible to steer innovations; one cannot elicit innovation by legal fiat. Consequently, scepticism towards any kind of steering is widespread among evolutionary economists. Although we agree with these arguments in principle, we wish to argue in favour of a different position. Steering is possible, but the philosophy of steering innovation processes is considerably different from a "press the button and get a particular result" approach. One must opt for a modulation approach and engage in transition policies: moving away from undesirable solutions to better solutions and systems. Essentially, it is a coordination process in a context of uncertainty (creating a knowledge problem) and vested interests (creating a governability problem) [31].

In the second Section of this paper, the main findings are summarized that have been collected by evolutionary economics and innovation theory on the nature of technological change. Based on this, a different notion of ecological problems, compared to the one used in conventional textbooks, is developed in the third Section. The reader will easily see that this notion essentially is a dynamic one. An outline of environmental innovation policy is developed in the fourth Section. We mainly focus on some important features of such a policy in the sense of the philosophy mentioned above, knowing that there are still many open questions that cannot be discussed in the limited space of this paper. Nevertheless one problem which seems to be somewhat neglected in the present literature is addressed in more detail in the fifth Section; namely how conflicts, especially conflicts with lobbyists of incumbent industries, can be moderated. A few concluding remarks finish the paper.

2 Technological Change from an Evolutionary Economists' Point of View

In modern evolutionary economics the dynamics of technological change is described as an evolutionary process, i.e. an interplay between variations of technologies and selection processes [33]. The notion of variation stresses that evolution relies on past technology and institutions to a large extent, while selection means that from all the different variants (available or imaginable) some get selected, usually those which turn out to offer the best compromise in terms of performance characteristics and price or because of sheer luck, thanks to a first mover advantage. An example for this is the QWERTY keyboard. Which technology will win cannot be predicted for sure because of uncertainty about user needs and evolution of prices and costs.

The concept of technological change as a historic, path-dependent process with possibilities of "locked-in" development is worked out theoretically in Paul David's model of localised learning and Brian Arthur's model of increasing returns with adoption [1]. In the words of Paul David [7, p. 4]:

> "Because technological 'learning' depends upon the accumulation of actual production experience, short-sighted choices about what to produce, and especially about how to produce it using presently known methods, also in effect govern what subsequently comes to be learned. Choices of technique become the link through which prevailing economic conditions may influence the future dimensions of technological knowledge. This is not the only link imaginable. But it may be far more important historically than the rational, forward-looking responses of optimizing inventors and innovators which economists have been inclined to depict as responsible for the appearance of market- or demand-induced changes in the state of technology."

Economist Frank Hahn also notes that there is something essentially historical about economic equilibria:

> "The path of history is the outcome of individual decisions and in turn helps to fix the latter. This is really the main message: the information available to agents at any time is determined by the particular path followed. The economy could have followed a different path and generated quite different information in a proper definition of equilibrium and of course in the dynamics itself" (as quoted in [13, p. 4–5]).

If technological development is depicted as an evolutionary process in the sense sketched above the following questions must be answered: what are the items which are selected and, in our case, how should we define technology? What are the selection mechanisms for technologies? How does something new come into the world of technologies? And what are the dynamics of technological development according to time? Since a complete overview of the state of the art in evolutionary economics is not possible within the limits of this paper, the following paragraphs provide brief answers to these questions while mainly focussing on characteristics of technological development which are needed for the argumentation in the next sections.

In the literature, the notion of technology is used differently: as knowledge, an artifact, a socio-technical ensemble [4], or a configuration that works [40]. We simply define it as the body of knowledge which allows someone to manufacture a product or use it. This knowledge is contained in material technology, the skills necessary for its use (human capital), and the organisational way of combining the two.

The selection mechanisms include not only selection by product markets; selection is rather a multi-dimensional phenomenon [49, 51, chap. 1]. It includes selection of visions of future developments by capital markets. Selection takes place when new technologies must be adjusted to existing technologies with which they are combined. The existing infrastructure at a given point in time has a selective effect, private standards and public regulation work as a filter. Even social concern and political mechanisms have an impact in the sense of a selection process.

The dynamics of variation and selection of technologies is not a steady process. Phases of relative stability, in which a particular set of technologies often dominates a given techno-economic system, switch with phases of instability during which new technologies successfully overcome old ones. Generally, the main reason for stability in a techno-economic system are self-enforcing processes which lead to increasing returns of a dominant technology, which in turn betters its competitiveness compared to its younger rivals. Such self-enforcing processes can be brought about by economies of scale, economics of scope, network effects, learning effects, the advantages of specialisation, and division of labour. There are institutional sources of path dependency working simultaneously: i) vested interests in the continuation of a trajectory, ii) self-assumed roles of the actors, and iii) interpretative frameworks and beliefs. Each of these constitutes an important factor.

Therefore one can conclude that the development of a given techno-economic system is not completely flexible; normally it follows a trajectory (Dosi). As Dosi writes:

> "The patterns of technological change cannot be described as simple and flexible reactions to changes in market conditions: i) in spite of significant variations with regard to specific innovations, it seems that the directions of technical change are often defined by the state-of-the-art of the technologies already in use, ii) quite often it is the nature of technologies themselves that determines the range within which products and processes can adjust to changing economic conditions; and iii) it is generally the case that the probability of making technological advances in firms, organisations, and economies is, among other things, a function of the technological levels already achieved by them. In other words technical change is a *cumulative activity*" [original italics] [9, p. 223]

Despite the mechanisms favouring an existing trajectory, there have been changes in trajectory, both partial ones as e.g. with the move to digital photography and far-ranging changes as e.g. with the electrification of manufacturing and homes. One important reason for this appears to be that the problem-solving capacity of a dominant technology (the body of knowledge) is exhausted. In contrast to the arguments of Arthur [1] this has been worked out by Windrum following an argument of Frencken and Verbart [15]. Windrum writes that:

> "The functional form of the relationship between learning and the number of adopters is sigmoid. As the number of contributors increases in the initial phase of its history, so the problem-solving capacity of the user network supporting that technology increases exponentially due to gains of the division of labour and benefits from arising of new fields of application. However, there is an upper limit to the problem-solving capacity of a user network. As a technology paradigm matures, so co-ordination costs start to outweigh the gains derived through further division of labour (...). The ability to identify and develop new fields of application is similarly limited ..." [49, p. 302]

Therefore increasing returns of adoption are at some point bounded from above, a necessary, but not sufficient condition for technological change. According to [9], discontinuities in trajectories are associated with the emergence of a new paradigm. This happened with electrification and with the turbojet.

The heterogeneity of consumer preferences and markets can create market niches, which can be used for new technological ideas. Real markets (niches) are important because they facilitate processes of learning (about the technology and the market, social acceptance) and processes of societal embedding (capital formation, set up of distribution, dissemination of knowledge, adaptations at the user's side to facilitate the adoption, gain of user acceptance, removal of regulatory benefits, etc.) which are necessary for the further development of a new technology or technology system. They help to create virtuous cycles that allow a new technology to escape a lock-in by helping to overcome initial barriers of high costs, the non-availability (or high costs) of complementary technologies, and misfits between the new technology and the external environment during the infancy period of a new technology when it has not yet benefited from dynamic scale and learning economies [23].

For example, experiences with a new technology in the niche help to gain user acceptance, to alter established views and expectations (both on the supply and demand side), and to benefit from user feedback (about their needs and the functioning of a technology) which helps to determine companies' research, production, and marketing policies. As well, it helps to achieve cost economies in the production and use of the technologies, to promote the development of complementary assets, and to foster the building of a constituency behind a product, which is necessary for the exercise of political influence, the programming and pooling of research, or the introduction of quality assurance schemes. Niches thus provide an impetus to learning, investment, and alignment processes. The actual use of a new technology is crucial, as some things are only learned from experience [19][23]. Real experiences are often necessary for making institutional adaptations.

A good example of a process of niche development is the gas turbine, which developed from a supercharging device to an aircraft propulsion technology and from a propulsion technology to a technology generating heat and electricity, offering environmental benefits compared to steam turbines that constitute the dominant technology to generate electricity. The gas turbine thus developed into an "environmental technology" through a process of niche development [44]. Military demand has also often established a market for a new product [36]. Many radical technologies were first used for military purposes. There appears to be a need for creating niches for radical solutions offering environmental benefits. This is done for PV and may be done for fuel cells in the future [45].

It appears that niches are an important element in processes of co-evolution. Without a niche there will not be a mass market. They act as an incubator for a new technology, helping it to survive the selection pressures which are especially harsh for a new fledgling technology, and they act as a stepping stone for further change, for example the opening up of new areas of application and the development of a new regime in space and time.

Besides the strategic niche management, a further important concept for policy recommendations (see Sect. 4.1) is the notion of the "window of opportunity". This notion refers to phases of technological competition in which the techno-economic system in question is unstable due to its own dynamics. The concept has been used by various authors (such as [6, 38, 11, 25]), but has not been carefully defined. It refers to the temporary existence of circumstances that allow the creation of something novel. Using the notation of system dynamics we can define a window of opportunity as a time period in which the dynamic equilibrium is breakable with minor efforts. With respect to actors, we can characterise it as a time period in which innovative (economic, political or social) entrepreneurs have a particular chance to influence the long-term direction of technological, economic or social development more than during normal stable periods.

Based on the different kinds of technological competition we can also distinguish between two kinds of windows in the techno-economic system. Following the investigations of Kemp [25] and Reichel [39] concerning a competition which is dominated by the conflict between old and new technologies, we can call the first one the "Kemp-Reichel-window". This window is open if the investment cycle of old technologies comes to an end and new promising technologies are available at that time. Following the investigations of Arthur [1] and David [6] referring to new/new competition, old technologies no longer play an important role. We can refer to the second window as the "Arthur-David-window". This window comes into being in the early stage of competition between similarly developed technologies. Above all, this competition is decided by increasing returns to adoption. The direction of technological development may be strongly influenced by "small historical events" [1].

3 Ecological Problems as Development Traps

The present techno-economic systems of fossil-based energy, car-based mobility and industrialised farm production are not sustainable. Already now they are giving rise to serious problems. Although this fact is well-known, modern societies seem to have great difficulties in changing the direction of technological progress to be more sustainable. It is not very surprising that alternative systems of energy, transport and agriculture are increasingly becoming a target of environmental policy.

From a traditional point of view, the ecological problems described above are interpreted in economics as problems of internalisation. Since the external effects of unsustainable technologies are not internalised in the price system, the latter delivers misleading incentives, which favour unsustainable technologies and impede more sustainable ones. Additionally, with this type of incentive system entrepreneurs who develop environmentally friendly innovations do not receive a share of the social gains brought about by their innovations. The political implication of this theoretical framework is quite clear: it is the task of governments to rearrange the price signals in a more sustainable way.

What is most important, however, is that the distinction between internal and external loses its guiding function for policy in a dynamic context. A large part of modelling the internalisation of external effects and determining the prices for these effects is based on the assumption that firms and/or households make a choice between alternatives which have a constant economic and ecological performance over time, well known by the economic actors. More advanced approaches of modelling diffusion processes of technologies, such as Downing/White [10] or Milliman/Prince [32], assume that the feasible set of technologies is given and well-known and technology choices are only influenced by the price-performance ratio (for a more detailed analysis of these models see [22, part I] and [27]).

Barriers, which can restrict the feasible set of choices, are not included in such models, as well as technical progress which, in contrast, can enlarge the feasible set of choices. However, both, sunk costs and technological progress, can bring about reactions by politically induced internalisation through economic actors, who then display reactions different from those assumed in such models. Sunk costs can slow down the theoretical reactions; technological progress can accelerate them. In both of these cases the results will differ considerably from the results delivered in the model. Under these circumstances the welfare optimum will be a moving target depending on the more or less known velocity of technological progress.

Moreover, innovators are normally not certain, at least at the beginning of the innovation process, whether the outcome of the development process will be technically feasible, what the benefits will be in terms of profits and, what is important in our case, whether the innovation is truly an environmental success. Visions, experience, and learning are very closely connected with every innovation process. Even if many innovative firms built up a portfolio of innovations and reduced economic risk through sophisticated methods of handling uncertainty, it would not be possible to eliminate uncertainty completely. Innovation is and will always be an adventure. Moreover, from the governmental point of view, the social benefits of innovations are also fundamentally uncertain, especially for innovations supported for environmental reasons. The history of governmental support of technological development is full of examples of failed projects once held as great ideas. (It should be added that this is not an argument for abandoning public support for new technologies. It is rather an argument against the public expectation that public support of a particular technology is only justified if the government can guarantee its success.)

Even well known technologies can have surprising effects. At the beginning of the 20th century, people were concerned about cars. Many people believed that driving faster than horses or coaches was unhealthy and dangerous. Additionally, cars were noisy and many people believed cars should be banned for this reason. At that time, almost nobody could imagine that fossil fuelled cars would be a main reason for a phenomenon such as climate change. Although some scientists had a faint idea that burning of fossils could be a problem, this idea only gained prominence in the 1980ies. Another example are CFCs used as propellants and refrigeration agents. As a cheap, non-flammable, and stable agent they were viewed as perfect for use. It was discovered only 50 years after their invention that they destroyed ozone in the ozone layer.

Generally, every type of development presents a set of problems. No type of development will ever be sustainable in the sense that it can be continued forever without any kind of problem. As Nelson and Winter wrote in 1982 (long before Beck 1997) in their book *An evolutionary theory of economic change*:

> "The processes of change are continually tossing up new 'externalities' that must be dealt with in some manner or other. In a regime in which technical advance is occurring and organizational structure is evolving in response to changing patterns of demand and supply, new non-market interactions that are not contained adequately by prevailing laws and policies are almost certain to appear, and old ones may disappear. Long-lasting chemical insecticides were not a problem eighty years ago. Horse manure polluted the cities but automotive emissions did not. The canonical 'externality' problem of evolutionary theory is the generation by new technologies of benefits and costs that old institutional structures ignore" [33, p. 368].

One might say that we should not be surprised by surprises. At the beginning of a technological path the ecological problems brought about by a particular technology are often not known or not well understood. Ongoing experience closely connected with the spread of a new technology throughout the economy may deliver more insights, yet very often the new technology has already been well established until public concern arises about its ecological effects. So, we can very often observe that, at the beginning of such a process, an internalisation of external effects is not a real alternative, simply because the effects are not known, not very well understood, or very disputed. It is also resisted by those who have to foot the bill of internalisation policies. In the end we have such a great amount of potentially vested interests created by the sunk costs of capital and jobs that an ambitious internalisation programme no longer seems to be politically feasible. This is the so-called "anticipation and control" dilemma about which Collingridge has written. External costs are not known at the time when possibilities for control are largest. In this sense, one can say that modern industrial societies are locked in many unsustainable paths of technological development or in development traps.

It is not just that firms keep consumers locked into an existing old technological system. The same is true for governments. Environmental and safety standards are usually based on well-proved compliance technologies, what hinders the adoption and development of more advanced technologies. Industrial policy is often aimed at the protection of old industries that are challenged by new firms and technological advances. Time is needed for new skills and ideas to penetrate the education system, and so on. The key problem for new technologies to become incorporated into the socio-economic system is that of compatibility. Within the process of economic development, technical interrelations and institutional rigidities have developed and may hinder technological shifts. New technologies which can easily be embedded in the production system and people's ways of life will diffuse more rapidly than technologies which require the replacement of capital goods, a new infrastructure, different skills, new ideas about production and consumption, and regulatory changes. Not only do the characteristics of the selection environment determine the relative use of technologies over time, but these characteristics also have implications on the kind of search activities which are likely to be undertaken by for-profit organizations.

The above-mentioned aspects help to explain why manufacturers often strive to develop so-called "drop-in" innovations which can easily be introduced in existing production processes and require few changes in the selection environment. For example, in the case of chlorofluorocarbons (CFCs), research efforts are directed towards the development of CFC substitutes (e.g. as cooling medium in refrigerators) that can easily be embedded in the economic and social environment rather than towards the development of totally different production techniques and products (e.g. a refrigerator with a totally different cooling system). Not only do the manufacturers of CFCs have an interest in developing these innovations which belong to the old CFC trajectory, but so do the users of CFCs [46].

Of course, there are good economic reasons for relying on drop-in solutions, but in so doing opportunities for system innovation are forgone. There is a need for policy not only in order to upgrade an existing system environmentally, but also to facilitate processes of system innovation which offer long-term benefit.

Thus the general target of an innovation oriented environmental policy is to overcome these development traps while preventing new ones. Such a policy must not only solve particular sustainability problems; it should improve the capabilities of policy in doing so. The enhancement of adaptive capabilities on the part of government or society is being called second order sustainability [43].

4 Suggestions for an Environmental Innovation Policy

Many evolutionary economists do not engage very much in policy recommendations due to the knowledge restriction policy must face. If the future is open, such an attempt seems to be futile and useless. Still, such a statement does not mean that the wind of change can blow in any imaginable direction. The notion of path dependency signals that some directions of change are more likely to occur than others. If technological development is path-dependent, the main economic, political, and cultural drivers of such a path dependency can be identified and statements will be possible whether a given economic system is more likely to change in an incremental way or to undergo fundamentally changes: what is possible is pattern prediction. Such a prediction not only includes statements about the stability features of a given techno-economic system; sometimes it is even possible to anticipate a window for major changes. Endogenous development and exogenous "shocks" from the broader selection environment can destabilise a techno-economic system and can create an opportunity for systemic innovations. If such an opportunity can be used by entrepreneurs, the outcome of this attempt is dependent on many contingencies and cannot be foreseen. The precise outcome of incremental technological change and the outcome of more fundamental changes cannot be predicted in advance. Insofar we follow the argument inspired by evolutionary economics that knowledge restrictions limit the capability of governments in steering innovation processes (and the capabilities of private actors as well); it is possible to say, however, that incremental or fundamental changes of a given technological system are more likely to occur in the future.

Against this background, an appropriate governmental steering philosophy is based on two elements. Different kinds of flexibility of techno-economic systems in the course of time have to be systematically integrated in the design of an environmental policy aiming at environmentally friendly innovations. This task is addressed in the "Sustime project" and will be described in the following first Subsection. Given knowledge restrictions, such a policy has to be adaptive and open for learning processes. This task is addressed in the concept of transition management and will be investigated in more detail in the second Subsection.

4.1 Preparing, Using, Opening, and Closing Windows of Opportunities

One can describe a transition process as a sequence, beginning with an old path and the discovery that this path is not sustainable and no promising solutions are available at the outset. The sequence ends when a transition is completed and market forces are reinforced. The important aspect is that policy strategies differ considerably according to the different stages of such a sequence. Using the term "window of opportunity" makes a difference for environmental policy whether policy comes about before the window is open, during the window and afterwards, when a transition is completed (for a more detailed discussion see [51, last chapter]).

Below, we describe various possible situations providing the stability features of a given techno-economic system in more detail.

1) The first situation can be characterised by three features: A sustainability problem linked to the old path is detected, the old path is *stable*, no techno-economic window exists, and *no promising solutions* are available. The main target to improve flexibility of the system is to stimulate the development of promising solutions, mainly by supporting scientific research and providing incentives for firms to adopt new scientific ideas.
2) If promising solutions are available, we can go on to the next step. The second situation is characterised by a *stable* old path, but now there is at least *one promising solution*. The main targets to improve second-order

No	Status of the techno-economic system	Type of competition	Quality of alternatives	*Strategies*
1	stable	not applicable	only theoretical alternatives exist	*Window preparation*
2	(still) stable	not applicable	promising solutions	*Window preparation*
2.a	stable	not applicable	promising solutions	*Window opening*
3	unstable	old vs new and new vs new	at least one competitive solution	*Window utilisation*
3.a	unstable	mainly new vs new	one alternative solution is competitive, there are other promising solutions	*Window utilisation*
3.b	unstable	mainly new vs new	multiple alternative solutions are competitive	*Window utilisation*
4	stable	not applicable	transition is completed	*"Window closing"*

Table 1: Strategic options according to system features

sustainability are to create diversity and to stimulate firms to develop at least one competitive solution, for example by organising learning curves. The government should make best use of market forces; here this involves mainly searching for new promising solutions and developing new solutions until they become competitive to some extent. For policies which *prepare* the emergence of techno-economic windows, expectation management is important. Weak signals, such as long-term targets, also might play a role. Mechanisms may e.g. include the creation of niches for, or the support of, new alternatives (strategic niche management). Additionally, we must keep in mind that environmental policy requirements may also hinder window emergence, e.g. by delay investment cycles (retrofitting), thereby increasing sunk costs, especially if end-of-pipe treatment is involved. In this case, transition might be obstructed by environmental policy itself.

2.a) A situation very similar to 2) arises when strong social or (international) political pressure forces the government *to open a window using political means* under the conditions that the old path is *stable* and only *promising solutions* are available. This situation is different to that described under 2), because government has to deal with strong opposite market forces. Although this may be necessary, we must be aware that the danger of add-on-technologies or retrofitting of existing technologies increases considerably, especially if governments use instruments that stimulate quick solutions. Governments have an incentive to do so if the political window of opportunity is shorter than the time period required for developing more fundamental alternative solutions. In addition to the targets mentioned in 2), the government must balance the social pressure for a quick solution, needed for political support, and the time period needed for more far-reaching solutions.

3) If one or more solutions become competitive to some extent, the next step can be taken. This situation may generally be characterised by the following features: the old path is *unstable*, or at least a techno-economic window can be anticipated, and there is competition between different new solutions. At least one of the new solutions is *competitive* in principle. In short, we face a combination of *new/new*-competition and *old/new*-competition. Fundamentally, a transition is now possible and the government's target might be to facilitate this transition, for example by abandoning discriminating mechanisms for the new solution. For appropriate policies which *take advantage of* or *utilise* an emerging old/new window ("Kemp-Reichel window"), a relatively small and perhaps temporary political impulse might be sufficient. The main political task is to grasp the situation and to have flexible and well measured-out instruments available to deal with the dynamics; standards may have an advantage here.

3.a) In some cases, the situation is more complicated than in 3): besides the competitive solution there may be other solutions which are merely promising and have not yet attained competitiveness. The development of their potential can be strongly impeded by simply following the target of transition. If some new solutions can use network effects and early economies of scale they can gain an advantage and cannot be overtaken by other promising solutions with a possibly greater potential. In other words, there is a trade-off between diversity and facilitated transition. In this situation, the government must keep the window open by suppressing the selection function of markets until the most promising solutions have developed their potential. If this is too costly, or not feasible and the old/new window can only be used by the more advanced technologies, a lock-in of new solutions must be avoided at least, e.g. through reservation of niches etc.

3.b) Sometimes the necessity of a transition is due to internal limits of the old path. As a result, new/new-dynamics come to the forefront. For policies which *take advantage of* or *utilise* these new-new techno-economic windows, "utilise" can also mean "keep the window open" for a sufficient time-period. Political responsibility is also high here: environmental policy may act as the "small historical event" within the selection environment which is important for the increasing returns models, e.g. biases competition. This may reinforce, or even lock-in, first mover advantages. The political exploitation of techno-economic new/new windows mainly consists in assuring that, in an open phase of competition, the best technologies in ecological and economic terms have the chance to be selected.

4) The key question after completed transition is whether gains of dynamic allocation efficiency can justify the losses of static allocation efficiency by suppressing the selection function of markets. If no further technolo-

gical progress of new technologies can be expected, the main target in this situation will be a proper selection of best solutions available. This sometimes means that government has to end all political interference in market processes. This step is important because subsidies, protected markets, and other political support create their own momentum; they bring about sunk costs and create many vested interests when support is ended.

Table 1 below provides an overview of the situation described above and the appropriate policy strategies.

4.2 An Integrated Approach: Transition Management

A learning-based adaptive approach is best undertaken as part of a broader transition approach. An example of such an approach is the model of transition management in the Netherlands. The goal of transition management for sustainability is to orient socio-technical and political dynamics to sustainability goals chosen by society. Transition management can be described as a forward-looking, adaptive, multi-actor type of governance aimed at long-term transformation processes offering sustainability benefits. One tries to steer processes of co-evolution in a reflective manner. Transition management is concerned with the functioning of the variation-selection-reproduction process: creating variety informed by visions of the sustainability, shaping new paths, and reflectively adapting existing institutional frameworks and regimes. The concept of transition management is used in the Netherlands as the model for sustainability policy. It has been developed by Rotmans and Kemp for the fourth National Environmental Policy Plan (NMP4) (The model is elaborated in [26, 8] and [27], c.f. also R. Kemp's contribution in the present volume). It is a model for escaping lock-ins and moving towards long-term solutions offering multiple benefits, not just for users, but also for society as a whole. It is motivated by broad social welfare considerations instead of environmental goals. In fact, economic considerations of creating new business through innovation play an important role in it.

A schematic view of transition management is given in Figure 1 above.

Embedding transition goals and policies in institutional arrangements is a key element. Transition management is thus not only concerned with technologies, but also with institutional change.

Policy actions are evaluated against two types of criteria: 1) the immediate contribution to policy goals (for example in terms of kilotons of CO_2 reduction and reduced vulnerability through climate change adaptation measures), and 2) the contribution of the policies to the overall transition process. This means that, under transition management, policies have a content goal and a process goal. Learning, maintaining variety, and institutional change are important policy aims which are used as means for change. The evaluation and

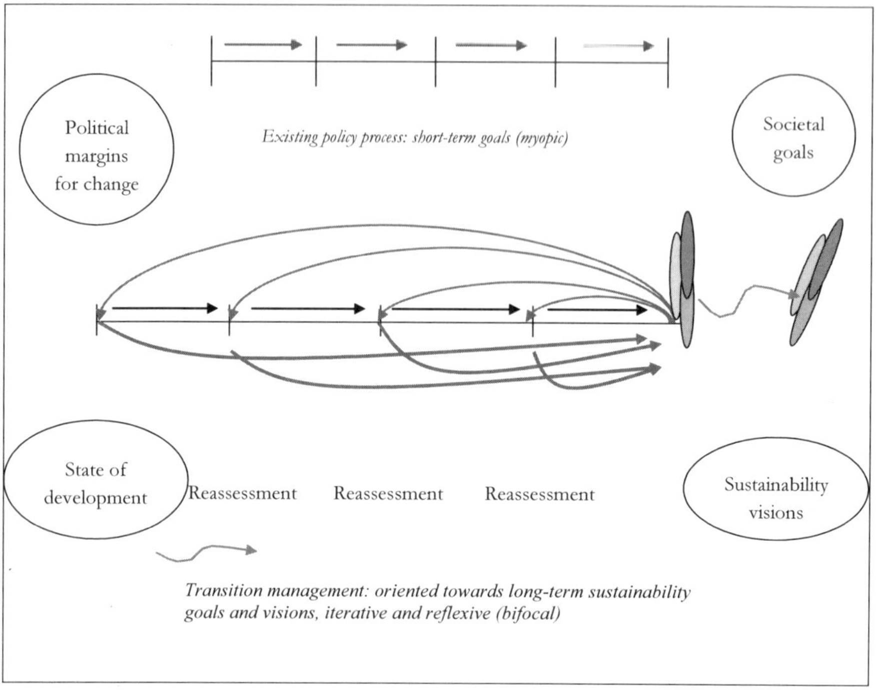

Fig. 1. Current policy vs. transition management

adaptation of policies (strategies, involved actors, progress, etc.) during development rounds brings flexibility to the process, without losing the long-term focus.

The role of government differs for each transition phase. For example, in the pre-development stages there is a special need for social experimentation and support for a transition programme, the details of which should evolve with experience. In the acceleration phase there is a special need for controlling the side effects of the large-scale application of new technologies. Throughout the entire transition the external costs of technologies (old and new ones) should be reflected in prices. This is not easy: taxes are disliked by anyone who must pay them. Perhaps it helps if taxes are introduced as part of a politically accepted transition endeavour, and if the revenues are used for funding the development of alternatives. Overall, transition management requires new roles and new modes of operation, especially for governments, which deal with the specific characteristics of transition processes. This means that a policy transition towards a more flexible, participative, and facilitating government is necessary.

Transition management does not dictate that one should achieve system innovations at all cost. It opts both for system improvement (improvement of an existing trajectory) and system innovation (representing a new trajectory of development or transformation). Transition management puts policies in a different, long-term perspective and tries to better align specific policies. The alignment of policy fields and new institutions of transition management is depicted in Figure 2.

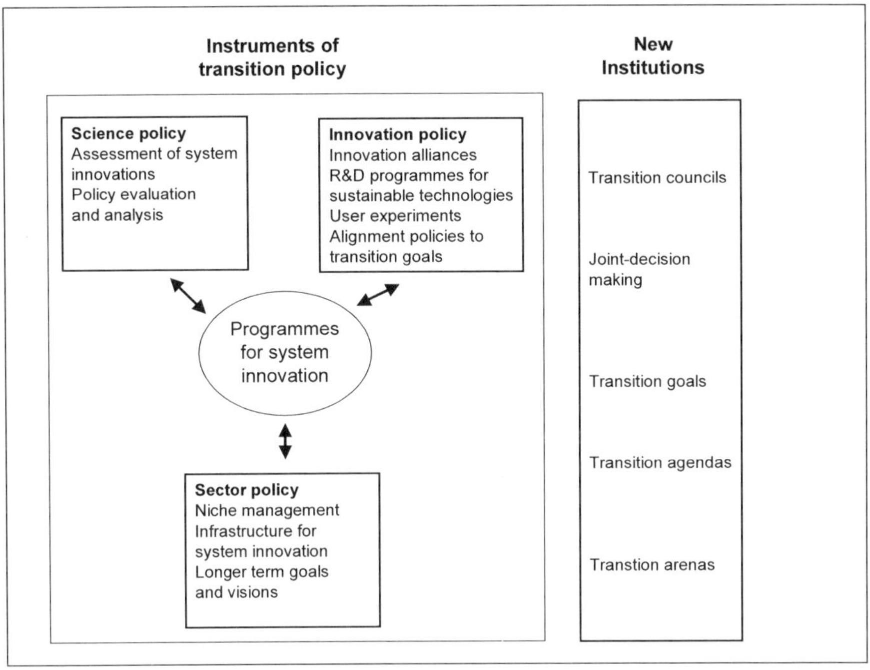

Fig. 2. Alignment of policies and instruments for transition management

Transition management goes beyond instrumental choices, as shown by the right-side box describing the new institutions connected with transition management. A central element are programmes for system innovation equipped in the course of time with exit strategies. Within these programmes, strategic experiments (see [23]) play an important role.

In managing transitions several aspects require special attention [27]:

1. One should be careful not to get locked in suboptimal solutions. This calls for anticipation of outcomes and the use of markets for coordination and context control instead of planning. A second way of circumventing lock-ins is by exploring different configurations through *portfolio management* – a common strategy in finance to hedge risks. One should not bet on one horse only but explore a wide variety of options, both incremental and radical ones.

2. One should *embed* transition policy into existing decision-making frameworks and legitimise transition management. Transition management should be politically accepted and be a joint concern for different policy makers and society at large. Long-term goals chosen by society should guide policy in addition to short-term concerns.
3. One should *ascertain a dynamic mechanism of change*, making sure that the process does not come to a halt when positive results fail to materialise immediately due to setbacks. One way of doing so is by making learning a policy objective.
4. One should engage in *multi-level coordination*: coordinate top-down policies with bottom-up initiatives (engage in vertical coordination besides horizontal coordination). Experience from local experiments should be shared for policy making on the national level and there should be strategic experimentation for system innovation, two things that have not happened in the past. There should be more and better coordination between top-level and local policies and also between various horizontal policies. National strategies should be coordinated with international policies because go-it-alone initiatives can be harmful unless there are clear first-mover advantages.

In the Netherlands, the national government committed itself to transitions toward sustainability in energy, transport, water management, and agriculture. For this it is using the model of transition management, which demonstrates that it is not just a theoretical fancy. Transition management is not simply an instrument, but rather a perspective. It is not based on blueprint thinking. No choice is made as to future functional systems. Different visions and routes are investigated through adaptive policies: decisions are made in an iterative way and support is temporary, which means that there should be "exit strategies".

A last point to be considered regarding the transition process is fairness. Many of the most desirable sustainability-oriented initiatives will involve trade-offs, including inequitable distribution of gains and losses. Such inequities are particularly worrisome where the losses threaten to be suffered by those who are already disadvantaged (a sadly common feature of past development assistance projects). Preparing for just transitions [5] which avoid disadvantages and provide satisfactory compensation when everything else fails is crucial. This problem is addressed in more detail in the following Subsection.

5 How to Handle Conflicts

Transitions are usually not an easy and pleasant game. Win-win-situations are rare. In fact, costs and benefits are usually inequitably distributed among the involved members of society. Especially actors who are believed to mainly bear the costs have strong incentives to obstruct change and engage in rent-seeking activities. Therefore the conflict management has to be an integrated part of an environmental innovation policy and concepts such as transition management.

Usually in environmental economics, allocation and distribution problems are conceptualised as a two-stage process [50]. Firstly, an optimal allocation has to be identified and implemented by a reconfiguration of property rights by the government. Then, in a second stage, upcoming distribution conflicts must be solved by governmental transfers to social groups that are becoming more or less worse off through the process of economic change. Unfortunately, in the political arena, allocation and distribution problems are closely linked and must be managed simultaneously in order to receive a solution to transition problems. The bargaining involved in transition goals, strategies and instruments is in almost every case an attempt to simultanously solve allocation and distribution problems [3, 47]. If the intended solution is not perceived as a fair burden sharing, political success is not likely to occur. Against this backdrop the handling of distribution conflicts is an important condition for a successful transition management.

There are a few interesting starting points for moderating distribution conflicts, if such conflicts are seen as a dynamic phenomenon for itself. Some of them are briefly outlined below:

- Transition processes need time. A well-defined goal in combination with a moderate and flexible time schedule allow the involved actors, especially the firms, to look for and select individual transition strategies minimizing the burden of transition costs. A synchronisation with the investment cycle might then be possible so that losses in terms of capital and jobs are lowered.
- Innovations which build a bridge between old an new trajectories might also be helpful, not only for the promoters of new technologies, but also for actors involved in the traditional technological trajectories allowing them an easier engagement in the new development.

Many case studies show the overwhelming importance of the existence of a promising solution [51]. If a transition seems to be impossible because no technically and economically feasible solution is at hand, most actors tend to defend the status quo, since the economic risks of a transition are supposed to be too high and the benefits cannot be foreseen adequately. In contrast, if a transition is perceived as technically and economically feasible because a promising solution exists, we often observe a switch in the public belief system about the risks, costs, and benefits; thus transition might be easier. Based on these empirical findings, a preparation of time windows throughout the creation of variety of possible solutions is very important. It is important to not simply go for the most economic option at any time, but to nurture long-term options besides short-term ones. Policy should not only nourish new options but also deal with the negative side-effects.

This consideration leads to the phenomenon of co-evolution of belief systems and technological change. Risks, costs, and benefits of technologies cannot be anticipated completely; they are always subject to learning processes. Due to the lack of "objective" information, technological development is, to a great extent, vision-based, relying on a framework of beliefs which are used to estimate current and future chances, risks, costs, and benefits.

Besides culturally embedded risk attitudes, which vary considerably between different cultures of capitalism, there is obviously a structural bias favouring mature technologies. The knowledge base is usually better for old technologies than for new ones because of the learning processes in the past. Moreover, such belief systems often have a strong conservative momentum since generating knowledge is not free of costs. Single economic actors and interest groups therefore act on the basis of routines, at least in the beginning of bargaining processes regarding transitions. This is very much in line with the idea of Schumpeter that an entrepreneur will see the tableau of opportunities, with their costs and benefits much differently than the pure view of economic common sense. For that reason, communication is in itself a resource helping to overcome routines and distribution conflicts.

Against this backdrop, careful monitoring of technological opportunities, which sometimes breaks up traditional beliefs, might be a first step in solving distribution conflicts. The second step could involve a switch in the framework of common beliefs. Similar to Thomas Kuhn's idea that the history of science can be described as a succession of scientific paradigms, fundamental technological changes are almost in every case accompanied by a considerable change in technological visions. Consequently, the assessment of risks and chances, costs and benefits also changes if embedded in a new belief system. This is why the role of learning processes is very much emphasised in the concept of transition management.

6 Concluding Remarks

Steering requires information about causal links and possible effects. Political actors often do not possess information on the effects of instruments for use. The widespread opinion in the relevant literature is that political actors are not well informed about many important features of technological development, and we agree with this point. Obviously, such a knowledge base for policy is far beyond reach. However, what is possible is pattern prediction in the sense of Hayek [18]; what is certainly impossible are predictions of the outcome of technological development. Mainly for that reason the approaches of time strategies and transition management must be understood in the sense of guiding lines by which political action under the condition of uncertainty should be measured.

One last remark could be helpful for understanding the real problem policy must face. By emphasising the limits of knowledge of political actors, many scholars allege that political actors are free in choosing a generic or selective approach of technology or environmental policy and that they should choose a generic approach since such a policy is far less demanding based on knowledge limits. Due to path dependency, however, even generic measures such as taxes or tradable permits often end up being selective depending on the time of implementation. In a stable phase generic measures mainly bring about further improvement within the limits of dominant technologies; in an unstable phase generic approaches can – but do not have to – create more fundamental technological changes. Taking this for granted, political actors often do not have a real choice between a generic and a selective approach. Empirical findings bear out this claim to some extent: almost every regulation scheme has a technological content discriminating against some technologies and supporting others. In the light of this background, the actual question is how far political actors should – and can – improve their knowledge base while admitting that they face severe restrictions in doing so.

A good timing of policy strategies according to the features of dynamic systems, an appropriate dosage of political measures, and a learning-based adaptive approach developed by the concept of transition management offer solutions for the knowledge problem with which environmental policy is confronted.

References

1. Arthur W.B. (1989): "Competing Technologies, Increasing Returns, and Lock-in by Historical Events". Economic Journal, 99, 116–131
2. Beck U. (1997): "The Reinvention of Politics: Rethinking Modernity in the Global Social Order". Polity Press
3. Beckenbach F. (1992): "Ökologisch-Ökonomische Verteilungskonflikte" (Diskussionspapiere des IÖW). Berlin, Institut für Ökologische Wirtschaftsforschung
4. Bijker W.E. (1995): "Of Bicycles, Bakelites, and Bulbs. Toward a Theory of Sociotechnical Change". Cambridge, MA, MIT Press
5. Burrows M. (2001): "Just Transition,". Alternatives Journal 27.1
6. David P.A. (1985): "Clio and the Economics of Qwerty". In: American Economic Review, Vol. 75 (2), 332–337
7. David P.A. (1987): "Some new Standards for the Economics of Standardization in the Information Age". In: Dasgupta P. and Stoneman P. (eds.): Economic policy and technological performance. Cambridge, MA, 206–239
8. Dirven J., Rotmans J. and Verkaik A.-P. (2002): "Samenleving in Transitie. Een Vernieuwend Gezichtspunt". LNV, ICIS en Innovatienetwerk Groene Ruimte en Agrocluster, April 2002
9. Dosi G. (1988): "The Nature of the Innovation Process". In: Dosi G., Freeman C., Nelson R.R., Silverberg G. and Soete L.L.G. (eds.): Technical Change and Economic Theory. Pinter Publishers, London, New York
10. Downing P.B. and White L.J. (1986): "Innovation in Pollution Control". Journal of Environmental Economics and Management, 13, 18–29
11. Erdmann G. (1993): "Elemente einer evolutorischen Innovationstheorie". Tübingen, Mohr Paul Siebeck

12. Freeman C. (1984): "Prometheus Unbound". Futures 16 (5), 494–507
13. Freeman C. and Perez C. (1988): "Structural Crises of Adjustment, Business Cycles and Investment Behaviour". In: Dosi G., Freeman C., Nelson R.R., Silverberg G. and Soete L. (eds.): Technical Change and Economic Theory. London, New York, Pinter Publishers
14. Freeman C. and Soete L. (1997): "The Economics of Industrial Innovation (Third Edition)". Pinter, London
15. Frencken K. and Verbart O. (1998): "Simulating Paradigm Shifts Using a Lock-in Model". In: Ahrweiler P. and Gilbert N. (eds.): Computer Simulations in Science and Technology, Berlin, Springer Verlag
16. Geels F.W. (2002): "Technological Transitions as Evolutionary Reconfiguration Processes: A Multi-level Perspective and a Case-study". Research Policy, 31(8/9), 1257–1274
17. Grunwald A. (2000): "Technology Policy between Long-term Planning Requirements and Short-ranged Acceptance Problems. New Challenges for Technology Assessment". In: Grin J. and Grunwald A. (eds.): Vision Assessment: Shaping Technology in the 21st Century Society. Towards a Repertoire for Technology assessment. Berlin-Heidelberg, Springer, 99–148
18. Hayek F.A. (1969): "Freiburger Studien". Tübingen
19. Hoogma R., Kemp R., Schot J. and Truffler B. (2002): "Experimenting for Sustainable Transport: The Approach of Strategic Niche Management". London, SPON Press
20. Kemp R. and Soete L. (1992): "The Greening of Technological Progress: An Evolutionary Perspective". Futures 24(5), 437–457
21. Kemp R. (1994): "Technology and the Transition to Environmental Sustainability. The Problem of Technological Regime Shifts". Futures 26(10), 1023–46
22. Kemp R. (1997): "Environmental Policy and Technical Change. A Comparison of the Technological Impact of Policy Instruments". Cheltenham, Edward Elgar
23. Kemp R., Schot J., and Hoogma R. (1998): "Regime Shifts to Sustainability through Processes of Niche Formation. The Approach of Strategic Niche Management". Technology Analysis and Strategic Management, 10(2), 175-195
24. Kemp R. (2000): "Technology and Environmental Policy. Innovation Effects of Past Policies and Suggestions for Improvement". OECD proceedings Innovation and the Environment, Paris, OECD, 35–61
25. Kemp R. (2001a): "Opportunities for a Green Industrial Policy from an Evolutionary Technology Perspective". In: Binder M., Jänicke M. and Petschow U. (eds.): Green Industrial Restructuring, Berlin etc., Springer, 151–169
26. Kemp R. and Rotmans J. (2005): "The Management of the Co-Evolution of Technical, Environmental, and Social Systems". In: Weber M. and Hemmelskamp J. (eds.) (2005): Towards Environmental Innovation Systems, Heidelberg/New York, Springer Verlag, 33–35
27. Kemp R. and Loorbach D. (2003): "Governance for Sustainability Through Transition Management". Paper for EAEPE 2003 Conference, November 7–10, 2003, Maastricht, The Netherlands
28. Kenny M. and Meadowcroft J. (1997): "Planning Sustainability". London and NY, Routledge
29. Lee K.N. (1993): "Compass and Gyroscope. Integrating Science and Politics for the Environment". Washington D.C., Island Press
30. March J.G. and Olsen J.P. (1995): "Democratic Governance". The Free Press, NY

31. Mayntz R. (1994): "Governing Failures and the Problem of Governability: Some Comments on a Theoretical Paradigm". In: Kooiman J. (ed.): Modern Governance. New Government-Society Interactions, Sage, London, 9–20
32. Milliman S. and Prince R. (1989): "Firm Incentives to Promote Technological Change in Pollution Control". Journal of Environmental Economics and Management, 17, 247–265
33. Nelson R.R. and Winter S.G. (1982): "An Evolutionary Theory of Economic Change". Cambridge, Bellknap Press of Harvard University
34. Nelson R.R. (1987): "Understanding Technical Change as an Evolutionary Process". Amsterdam, North Holland
35. Nill J. (2002): "Wann benötigt Umwelt(innovations)politik politische Zeitfenster? Zur Fruchtbarkeit und Anwendbarkeit von Kingdons "policy window"-Konzept". (Diskussionspapiere des IÖW 54/02), Berlin, Institut für Ökologische Wirtschaftsforschung
36. Nill J. (2005): "Techno-Political Competition and Lock-in: The Case of Nuclear Power Technologies". In: Zundel S. and Sartorius C. (eds.) (2005): Time Strategies for Innovation Policy Towards Sustainability. Cheltenham (UK), Edward Elgar
37. NMP-4 (2000): "Een Wereld en een Wil. Werken aan Duurzaamheid". (A World and a Will. Working towards Sustainability), The Hague
38. Perez C. and Soete L. (1988): "Catching up in Technology: Entry Barriers and Windows of Opportunity". In: Dosi G., Freeman C., Nelson R., Silverberg G. and Soete L. (eds.): Technical Change and Economic Theory. London, Pinter Publishers, 458–95
39. Reichel M. (1998): "Markteinführung von erneuerbarer Energie. Lock-out Effekte und innovationspolitische Konsequenzen für elektrische Wind- und Solarenergienutzung". Wiesbaden, Deutscher Univ.-Verlag
40. Rip A. and Kemp R. (1998): "Technological Change". In: Rayner S. and Malone L. (eds.): Human Choice and Climate Change, Vol 2 Resources and Technology. Batelle Press, Washington D.C., 327–399
41. Rotmans J., Kemp R., van Asselt M., Geels F., Verbong G., and Molendijk K. (2000): "Transities & Transitiemanagement. De casus van een emissiearme energievoorziening". Final report of study "Transitions and Transition management" for the 4th National Environmental Policy Plan (NMP-4) of the Netherlands, October 2000, ICIS & MERIT, Maastricht
42. Rotmans J., Kemp R., and van Asselt M. (2001): "More Evolution than Revolution. Transition Management in Public Policy". Foresight 3(1), 15–31
43. Sartorius C. (2003): "Second-order Sustainability – the Conditions for a Sustainable Technology Development in a Dynamic Environment". Ecological Economics 58/2 (2006), pp 268-286
44. Sartorius C. (2005a): "Combined-cycle Gas Turbines – Between Climate Protection and other Policy Objectives". In: Zundel S. and Sartorius C. (2005) (eds.): Time Strategies for Innovation Policy Towards Sustainability. Cheltenham (UK), Edward Elgar
45. Sartorius C. (2005b): "Stationary Fuel Cells and the Decentralized Cogeneration of Power and Heat". In: Zundel S. and Sartorius C. (eds.) (2005): Time Strategies for Innovation Policy Towards Sustainability. Cheltenham (UK), Edward Elgar

46. Sartorius C. (2005c): "Phase-Out of CFCs and the Protection of the Ozone Layer". In: Zundel S. and Sartorius C. (eds.) (2005): Time Strategies for Innovation Policy Towards Sustainability. Cheltenham (UK), Edward Elgar
47. Soete L. and Zundel S. (1993): "Ökologische Verteilungskonflikte". Literaturstudie im Auftrag der Hans-Böckler-Stiftung
48. Riele T. et al. (2000): "Transities: kunnen drie Mensen de Wereld Doen Omslaan?". STORRM, Twynstra Gudde Management, and Hötte Milieu Management
49. Windrum P. (2003): "Unlocking a Lock-in: Towards a Model of Technological Succession". In: Saviotti P.P.: Applied Evolutionary Economics – New Empirical Methods and Simulation Techniques. Cheltenham (UK), Northampton US., Edward Elgar, 292–321
50. Zimmermann K. (1986): "Discussion: Distributional Considerations and the Environmental Policy Process". In: Schnaiberg et al. (ed.): Distributional Conflicts in Environmental Resource Policy, 95–108
51. Zundel S. and Sartorius C. (eds.) (2005): "Time Strategies for Innovation Policy Towards Sustainability". Cheltenham (UK), Edward Elgar

Comment: Moderating Instead of Steering?

Frank Beckenbach

University of Kassel, Faculty of Economics, Nora-Platiel-Str. 4, D-34127 Kassel.
beckenbach@wirtschaft.uni-kassel.de

In the article of Kemp and Zundel a lot of valuable insights of innovation and environmental economics are combined and synthesized for specifying the possibilities as well as the constraints for political regulation of environmentally benign innovation processes.

The main message of the article can be summarized in the following quotation: "Steering is possible, but the philosophy of steering innovation processes is considerably different from a 'press the button and get a particular result approach"'(p. 25). The authors try to make this assumption plausible by dealing with three interrelated topics: the findings of evolutionary economics and innovation economics about the structural patterns of the innovation process (I), the analysis of the dynamic nature of environmental problems and the corresponding difficulties for internalising negative externalities (II), and conceptualising a dynamic innovation policy resulting from these structures and problems (III). In the following I will comment on each topic separately and finally draw some general conclusions (IV).

1 Evolutionary and Innovation Economic Analysis of the Structural Patterns of the Innovation Process

In evolutionary economics as well as in innovation economics the innovation process is analysed primarily from a meso, or even a macro perspective. Here the main focus is on cumulative interaction effects resulting from the relationship between the time of practicing a novelty and the number of its adopters. Especially the comparative advantages for an increasing number of adopters due to economies of scale and/or scope, learning effects and network effects are at the centre of analysis. These comparative advantages in the take-off situation of a new option (technology) are the reason why the development of new technologies is considered as a reinforced path-dependent process. The authors claim that this path-dependency is accomplished by cognitive framing effects as well as by activities for protecting the expanding novelty. According to the initial difficulties of establishing an innovation, to the above mentioned dynamics after a take-off, and to a saturation effect in learning and network-building as well as in economies of scale and scope a sigmoid curve for the process of adopting an innovation (diffusion curve) is assumed (and often empirically proved) as a standard case. Given such a trajectory of a successful novelty diffusion – at least after take-off – it seems difficult (if not impossible) for a novelty in a less developed stadium to compete against a more adopted novelty. The economic advantage of the latter is proportional to the longer time it has been practised and to the larger number of its adopters.

Because Kemp and Zundel share these views of evolutionary economics about path-dependency they consequently ask how it is possible at all that an upcoming novelty can be successful (possibility question). Their answer to this question is that such a novelty needs the protection of a 'niche' for being successful. Here the notion of niche is not used in the sense it has in ecology where it refers to the capabilities of species to specialize according to the multi-facetted nature of environmental conditions. Rather a niche is meant here as a temporary protection against market competition by political authorities, or private consortia (cf. Kemp et al. 2001, p. 275). What is developed in the niche is either a solution for a special political (public) purpose complementary to the existing (private) options, or it is something which is compelling the established (private) options. In the first case a transfer of features of the niche product to other applications is required; in the second case a strong protection against competing established options is needed. Since both variants of niche management will not work without strong political assistance the answer to the possibility question mentioned above is tantamount to asserting that only by introducing exogenous forces a competition between different path developments seems possible!

My objection against this interpretation is twofold: *Firstly*, the authors follow the unconvincing tradition of evolutionary economics in identifying opportunities of cost reduction and/or productivity increase during path-dependent processes in a strategic meso-level perspective with that what actors can perceive and practise. In other words, it is assumed that actors always have a clear understanding of path-dependent advantages and that they have no other reason to leave the path. Both is not necessarily the case: There is no guarantee that boundedly rational actors can always anticipate the advantages of sticking to an option in terms of economies of scope, learning and network effects (e.g. the actor may wrongly anticipate saturation effects). Beyond a "press the button and get a particular result" perspective one has to take into account the internal conditions of actors in terms of their individual experience and absorptive capacities.[1] Hence, depending on his individual conditions an actor might be either forced or willing to search for a new option (cf. [3]).

Secondly: The cumulative effects defining path-dependency are a feature of the 'post revelation' phase of the novelty creation, i.e. the novelty is already there and it is in a more or less take-off situation. If the perspective is broadened to include 'pre-revelation' processes path-dependency is always threatened by the outcomes of invention having its own logic of generating, exploring new paths along "divergent thinking", and breaking with established frames of doing things (cf. [5, 1]). So there is always a stream of inventive ideas and concepts some of which might look promising for an entrepreneur leading to a strategic commitment of the latter to promote this new idea or concept.[2]

To resume, including the perspective of individual actor and the pre-revelation phase of novelty creation reduces the restricting implications which path-dependency has for the creation of new paths.

2 The Analysis of the Dynamic Nature of Environmental Problems and Difficulties for Internalizing Externalities

The impacts of the innovation process as a whole on the environmental conditions are not calculable in advance. The main reason for this is that at the beginning of an innovation path the direction as well as the cumulative dynamics of that path cannot be anticipated. Taking additionally the problem of sunk costs and vested interests into account – as the authors do – this leads to a problem of dynamic negative externality which is difficult to solve by internalisation: at the point of time when the negative (environmental) impacts of an innovation are finally known inducing a change by well known internalisation procedures is blocked by sunk costs, switching costs and the corresponding activities of vested interests. This is what the authors (following Collingridge) call the "anticipation and control" dilemma (p. 32).

Although the authors assert " that the distinction between internal/external loses its guiding function for policy in a dynamic context" (p. 30) they still seem to accept the externality/internalisation frame work of welfare theory as a useful concept in that dynamic context.[3] At least for three reasons one

[1] Due either to bad internal allocation of external information, or due to unsatisfycing experience when following a given path an agent might be willing to follow another path even if there still might be good exploitation opportunities for a given path.

[2] Such a commitment can be backed by a divergence between publicly articulated social needs and the outcomes of an established path.

[3] Otherwise there would be no "anticipation and control" dilemma! See also the quotation of Nelson/Winter on p. 32 where a dynamic welfare consideration is postulated.

might be sceptical about such a frame work for dealing with dynamic ecological problems. *Firstly*, the temporal and spatial dynamics of ecological systems principally prohibit to force the polluters to pay in many cases. Due to systemic complexities (like incubation processes, threshold effects, synergisms, and the multi-scale property of ecological systems) it is often impossible to find a clear cut functional relationship between emissions of identifiable economic activities and the corresponding impact of observable immissions. Then a systemic burden (like climate change impacts) and a systemic cause (like energy system or life style) have to be related to each other. This goes beyond the scope of (at least traditional) welfare theory. *Secondly*, there are methodological reasons for abandoning the welfare theoretic framework in an evolutionary context. An assumption made by almost all evolutionary economists is that the knowledge of economic actors is local according to their experience and perception capabilities. This excludes any sort of complete functional internal valuation scheme (either in terms of a complete utility function, or in terms of a complete production function and cost function) which is required for applying the usual internalisation devices. Furthermore, whatever the valuation scheme may be, it will change over time if learning is seen as an essential feature for an evolving economy.[4] Then the commensurability between the activities producing externalities and activities being harmed by the externality impact (which is required for a rational internalisation) is not given any more. *Finally* it is naive to think that in a world of complex ecological-economic interactions there could be political agents being guided by an a priori given welfare measure. Even if the latter would exist – what its improving means (let alone what its maximisation means) would then be a contested terrain.

3 Conceptualising a Dynamic Innovation Policy

Manifold conclusions can be drawn from discussing the patterns of economic innovation and the dynamics of environmental problems. (i) The more the process of innovation is patterned in an inflexible way (i.e. the more this process is path-dependent) the more restricted is the potential for political actors to influence this process. (ii) The possibility to use this potential is itself restricted by the fact that the future development of the innovation dynamics and the environmental impact resulting from it cannot be anticipated. (iii) Finally it is obvious that the informational restrictions as well as the complexity of the ecological-economic interactions exclude a traditional welfare perspective if environmental policy has to be conceptualised in such a frame work.

Conclusion (iii) is not explicitly drawn by the authors. Implicitly they seem to substitute 'sustainability' for 'welfare' as the overarching target for

[4] Cf. e.g. p. 41–42 where cost and benefits of technologies are considered as components of a learning process.

environmental policy. It is simply assumed that such a perspective is relevant for political actors and that there is a sustainability problem to solve in that the given modern market economies cannot be qualified as being sustainable (e.g. p. 30). But even if a consensus about sustainability should be possible in the future there are essential differences of such a policy orientation and a (traditional) welfare policy: the former is multi-dimensional (ecological, economic and social), it has a procedural (process-dependent) nature and requires an assessment of systemic impact constraints in all mentioned dimensions. How such a policy can be implemented given the (remaining) inflexibility of path-dependent innovation processes and the informational constraints in a coevolving economic and ecological system is not discussed by the authors. Hence, the question how to conceptualise an environmentally oriented (or even more ambitious: sustainable) innovation policy remains unanswered.

Taking into account the lack of conceptual foundations for innovation policy as regards to 'welfare' as well as regards to 'sustainability', I would suggest a pragmatic approach to conceptualise such a policy. Generally neither the impartial role of political actors in welfare theoretic explanation of policy nor the partial power maximising orientation of political actors in modern approaches of political economy seem to be adequate for explaining the political processes although both are pinpointing elements of the latter. A synthesis of these contradicting views is possible if the political actors, processes and regulations are considered as a social subsystem consisting of strong internal relations (such as the legislating, judicial as well as executing operations and the included power enhancing operations of political actors) and weak external relations to other subsystems like e.g. the economy (given by economic and social requirements articulated as public needs). Therefore it seems promising to analyse the relationship between economy and policy in terms of system components being "near decomposable" [6], each showing the multi-scale property, i.e. parts of these subsystems operating on different scales in terms of time and space. Then the dynamics of the political system is shaped by internal goals (giving room for a moderate variant of the power maximiser of modern political economy) as well as by external goals (giving room for a moderate variant of the welfare maximiser of welfare theory) which have to compromise at every scale. As regards to innovation policy the external goals can be specified as corresponding to the following requirements:

- promoting basic research (being a public good),
- transferring the results of basic research to private institutions,
- overcoming critical-mass problems in the diffusion phase, and
- initiating networks and cooperation for promoting the innovation process in general.

As regards to environmental policy the following requirements are the source for defining the external political goals:

- setting incentives for an environmental sensitivity of basic research,

- overcoming structural scarcity of natural resources with strategic importance,
- dealing with environmental conflicts,
- safeguarding institutions, and – perhaps –
- promoting sustainability orientation as part of a social learning process.

Contrarily to the neglected problem of a conceptual foundation of (environmental) innovation policy the authors explicitly take into account the above mentioned inflexibility problem (i) and the information problems for political actors (ii) in an evolutionary setting. Despite the severe information and knowledge restrictions for political actors they assume that the driving forces for the (technological) path development are known to the former allowing for the predictions of patterns for that development (p. 32–34). Reconciling the inflexibility of established innovation processes and the sustainability requirements for policy brings about two tasks for political actors: preparing, opening and closing "windows of opportunity" and "transition management".

"Windows of opportunity" are given if the established unsustainable path of technological development becomes "unstable" (due to what?), a new (more sustainable) path is known, and a social and/or political pressure for switching to the new path is available. Obviously this is more than simply knowing the driving forces of the existing path and its future patterns admitted to political actors in the view of the authors. Therefore the following questions arise: Can the "window of opportunity" be anticipated given the knowledge constraints of political actors? Can the knowledge about the "window of opportunity" (i.e. about the factors of instability of the existing path, about the new (more sustainable) path) be transformed into political regulation? What kind of instruments are appropriate for that? Can the effect of this mix of instruments be known in advance? How is such an idea about identifying and using a "window of opportunity" compatible with the "anticipation and control dilemma" mentioned above?

Even more ambitious seems to be the proposed "transition management": "Transition management is concerned with the functioning of the variation-selection-reproduction process: creating variety, informed by visions of sustainability, shaping new paths, and reflexively adapting existing institutional frameworks and regimes" (p. 37). Here the policy perspective is not confined by given "windows of opportunity", rather the aim is to create these windows. The core of such a transition management is establishing and developing a niche (cf. above). Obviously such a transition management is the task of a social planner evaluating policy actions, taking care that "... external costs of technologies (old and new ones) should be reflected in prices" (p. 38) and driving the whole process to a "system improvement" (p. 38). A lot of centralized information and knowledge is required for such a transition management and therefore it can be asked if such an optimistic view of transition possibilities is compatible with the evolutionary nature of the economic innovation process (discussed at length by the authors themselves) and any conceptualisation of the policy process based on the former. Taking the features of path developments into account it seems to me that a "steering" perspective is too optimistic[5] and should be substituted by a perspective in which the political actors are heterogeneous, but important moderators who can influence the intensity and the direction of the innovation process.[6]

4 General Conclusions

My general conclusion is that that Kemp and Zundel on one side overestimate the inflexibility of (technological) innovation processes because in analysing these innovation dynamics they follow the standards of evolutionary economics and innovation economics by ignoring the perspective of actors and the pre-revelation processes. Hence, to break up established paths is not the only possibility for an environmentally oriented innovation policy: another approach would be to strengthen and influence the pre-revelation processes of invention and early innovation. On the other side the authors underestimate the restrictions for establishing such a policy: there are severe problems of getting the necessary knowledge about future developments of technologies and its implications for the environmental conditions and there are constraints due to the internal logic of the policy process and its social embeddedness. I suppose that especially this point can be clarified if the standard approach to welfare theory is substituted by an evolutionary approach, emphasizing the process of social learning and consensus finding about the welfare goals, parting with the strict separation of allocation and distribution, and discussing intertemporal distribution conflicts (instead of externality/internalisation problems). Although some of these aspects are mentioned in the last section of the article they are not integrated in the analysis of the innovation process and the political advices (windows of opportunity, transition management) proposed by the authors.

References

1. Beckenbach F. and Daskalakis M. (2003): Invention and Innovation as Creative Problem Solving Activities – A Contribution to Evolutionary Microeconomics. Volkswirtschaftliche Diksussionsbeiträge, Universität Kassel
2. Kemp R. et al. (2001). In: Garud R. and Karnoe P. (eds.): Path Dependence and Creation. Lawrence Erlbaum Associates, Mahwah, NJ

[5] To rephrase it in the author's terms: There is a lot of buttons which have to be pressed and a variety of results which cannot be attributed to single buttons. Is that still 'steering'?

[6] Cf. the ex post cross country analysis of climate change and acid rain issues and the role of policy therein [4].

3. Lampel (2001). In: Garud R. and Karnoe P. (eds.): Path Dependence and Creation. Lawrence Erlbaum Associates, Mahwah, NJ
4. The Social Learning Group (2001): Learning to Manage the Global Environmental Risks: A Comparative History of Social Responses to Climate Change, Ozone Depletion, and Acid Rain.
5. Guilford J.P. (1959) Persönlichkeit: Logik, Methodik und Ergebnisse ihrer quantitativen Erforschung. Verlag Julius Beltz, Weinheim/Bergstr.
6. Simon H.A. (1996): The Sciences of the Artificial. 3rd edition, MIT Press, Cambridge, MA

Transition Management in the Electronics Industry Innovation System: Systems Innovation Towards Sustainability Needs a New Governance Portfolio

Joachim Hafkesbrink

Innowise research & consulting GmbH, Ludgeristrasse 20, D-47059 Duisburg.
jh@innowise.de

1 Introduction

In the Electronics Industry Innovation System (EIIS) a major shift from the current functional towards a more sustainable system can be observed. As a result of a new portfolio of different environmental policies, system innovations are expected to be taking place within the next two decades providing new opportunities for sustainability benefits compared to the present situation. It is expected that the highest sustainability potential in the EIIS is located in transition of the EIIS from a linear to a circular economy, covering a fundamental change in functional subsystems and product chains, actors configuration, business models, and so on. The transition stages are described in view of the interplay of the environmental policy portfolio developed over time, changes in the institutional framework of the value-ad chain, new business model paradigms, behavioural changes of consumers etc. as the main elements of the system innovation. The role of the government and other players of the innovation process is described in a multi-level analysis covering technology, production, user and policy regimes, giving examples of experimental niches (e.g. for new business models) and outlining overall setting in which processes of change occur. Furthermore, the implementation management of particular policy elements over time is depicted. Finally, recommendations are developed to backup the transition management in the EIIS towards a sustainable future.

This paper sketches selected results from the German BMBF funded research project "INVERSI" [1] under the umbrella of the program "RIW: Rahmenbedingungen für Innovationen zum nachhaltigen Wirtschaften" (Framework conditions for innovations sustainability) and from the European Thematic Network "ECOLIFE"[1], funded under the LIFE program of the Euro-

[1] See http://www.ihrt.tuwien.ac.at/sat/base/Ecolife/ECOIndex.html

pean Commission. It comprises case studies and empirical investigations as well as results from various expert workshops in these projects, especially directed towards an investigation of governance impacts on the Electronics Industry Innovation System.

The paper is organised as follows: In Chapter 2, problems and policy developments in the electronics industry are sketched which are leading to a transition of the innovation system. In Chapter 3, transition theory is introduced, and in Chapter 4 the developments of the EIIS are interpreted in terms of key elements of the transition theory (transition stages and multi-level aspects). Chapter 5 presents a discussion of possibilities for managing transitions and open questions.[2]

2 Problems and Policy Developments in the Electronics Industry

The electronics industry is regarded as a substantially dynamic innovation system. Due to rapid development of product innovations and shortening of innovation cycles, a broad variety of new electrical and electronic devices enters the market every year using, amongst others, also hazardous substances for particular functional features, as for instance flame retardands. Thus, in the EU in 1998 about 6,5 million tons of electronic waste with environmentally critical substances have been disposed, most of them via landfilling.

Since the total amount of Waste Electrical and Electronic Equipment (WEEE) generated in the EU is increasing by 16% to 28% every five years, both German waste management policy and EU environmental policy are following – with growing intensity in the course of time – the paradigm of the circular flow economy, and standardising (extended) producer responsibility (EPR) for the manufacturers or sellers of certain product groups.

The most important instrument they draw upon in the assignment of this producer responsibility is the take-back obligation aiming at avoiding waste, and increasing economic efficiency and ecological effectiveness of recycling and disposal. Thus, the disposal costs of products and packaging material shall be charged to the responsible producers and distributors. So, the producers shall be encouraged to consider the aspects of disposal as early as in the stages of design and production and to develop relevant innovations. After take-back regulations for packaging, batteries, and old vehicles were introduced during the last years, now the EC-directive for Waste Electrical and Electronic Equipment (WEEE directive) was passed on January the 27th 2003 which had to be transposed into national law until August the 13th 2004.

The WEEE directive marks a starting point for changes in the governance portfolio of the electronics industry leading to an ongoing transition of

[2] The author would like to thank all researchers and practitioners for their input, especially Kathrin Müller (Motorola), Gianlucca Brotto (Electrolux) and Prof. Ab Stevels (Philipps).

the whole innovation system towards sustainable development. Before we describe these transition processes and give some empirical findings, theoretical considerations are indispensable to understand the context of this paper.

3 Theoretical Considerations

3.1 The Sustainability Potential of System Innovations

In current literature, different types of innovations are discussed:

- incremental innovations, defined as innovations within a particular trajectory such as step-by-step improvements of a particular technology,
- radical innovations, i.e. ideas, not earlier known or used, driven to market success, connected with a quantitative and qualitative performance jump changing trajectories and knowledge paths, often as a combination of product, process, and organisational innovations[4], and
- system innovations, defined as fundamental changes of a system on a broad basis, accompanied by changes in multiple sectors of the innovation system; compound of incremental and radical innovations, the development of new actors configurations etc.[3]

System innovations involve changes in socio-technical systems beyond a change in (technical) components. They are associated with new linkages, new knowledge, different rules and roles, a new 'logic of appropriateness', and sometimes new organisations.[2] System innovations involve both a change in technology, products or services, and changes in market/actor configurations. The innovation types of incremental, radical and system innovation are depicted in Fig. 1, using some examples from the EIIS.[4] It is as well conjectured that the capability of generating sustainability benefits depends on the kind of system innovation, a result of combining scientific, technological, organisational, and structural changes regarding market and actors and, furthermore, may be critical with incremental innovations due to rebound effects. Thus, sustainable system innovations are defined as a particular kind of system innovation, comprising economic, ecological, and social improvements[4] as well as organisational, institutional, and even political elements influencing each other on the micro, meso and macro level.[5] The portfolio of sustainable innovation is depicted in Fig. 2:

As sketched in Fig. 2, the WEEE directive may be mapped as a political/institutional innovation comprising primarily economic and ecological issues. Corporate social responsibility programmes (CSR) may be indicated as organisational system innovations comprising economic, ecological and social concerns.

The sustainability potential of system innovations is assumed to be superior to incremental innovations and even radical innovations since it involves

[4] following the Brundlandt report

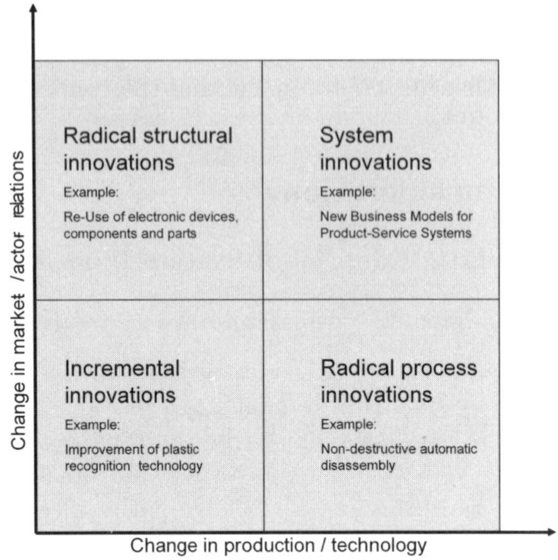

Fig. 1. Innovation types in the Electronic Industry[3]

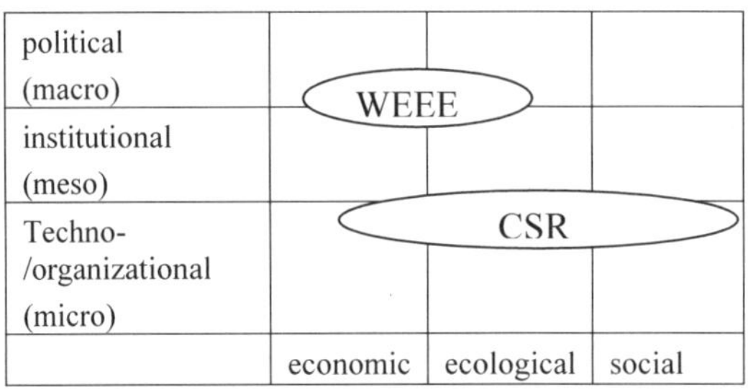

Fig. 2. Portfolio of sustainable system innovations using examples of the Electronics Industry

scientific-technical, organisational, and market shifts as well as changes in actor configurations. One of the most interesting aspects is the question, how to modulate the present system to enable a more or less endogenous collective transformation to reach a new level of sustainability by changing the innovation trajectory. System innovations for sustainability are almost always directed towards less resource-intensive regimes and are facing complex problems since they are expected to change the environmentally, and, in the long term, intergenerationally significant behavioural attitudes of different stakeholders of an innovation system. In this respect, the question arises, how to promote promising technological solutions to more sustainability, developed in niches, and help them to clear the hurdle of locked-in regimes on the meso level and to broadly diffuse on the macro level.

Against this background, the *main thesis of this paper* is that a shift towards sustainability via system innovations deserves a concept of transition management using a new governance portfolio consisting of particular macro, meso and micro level incentives. The concept of transition management is described in the following Chapter.

3.2 The Concept of Transition Management

This paper substantially draws on the concept of "transition management" as described by Kemp and Loorbach [2] with amplifications addressed by Berkhout, Smith, and Stirling [6]. Transition management is a new governance approach to overcome sustainability barriers like short term thinking, fragmented policies and institutional deficits, market imperfections externalising environmental costs, i.e. prices not reflecting the real costs of environmental degradation, violation of the polluter pays principle, great uncertainty of solutions, or insufficient precaution.[5] "The concept focuses on system innovations defined as a fundamental shift of technological, social, regulative, and cultural regimes, which in their interaction would satisfy distinctive needs like transportation, nutrition, housing, water, or energy. A system change in that sense involves a co-evolution of technologies, infrastructure, regulation, symbolic meanings, knowledge, industrial structures, etc. Such transitions typically take up a period of 30-40 years."[10]

The origin of "transition management" is to be located in the Netherlands, where most of the work on this issue in the research and policy arena as well as in interrelation with stakeholders takes place [11]. However, experience with transition management as well as theoretical work on this issue[6] provide a huge reservoir for explanatory and descriptive trials, for understanding practical problems of system innovation by applying theory elements and theses of transition management to particular empirical policy arenas and innovation systems outside the Netherlands. In this paper, the theoretical and heuristical potential of the transition management concept is applied to the EIIS with the attempt to mirror the empirical findings of the transition of this innovation system with theoretical insights of the transition management research.

In this respect this paper will draw on the theoretical framework, tackling in particular the following issues of transition and transition management:

(1) Transition management "has been defined as an anticipatory form of multi-level governance that uses collective, normative visions as starting

[5] [2], p. 4; in that sense the research on transition management at first glance has its parallels to former recommendations to set up for new enviromental policy styles [7] to introduce new steering principles [8] as well as to set up formal and informal institutiones in environmental policy [9].

[6] See [2], [5] and therein cited sources.

point for formulating long-term, collective innovation strategies ... This managerial approach advocates an evolutionary way of steering instead of command-and-control governance. It suggests that a transition takes place through a sequence of the following stages: *a pre-development phase* where there is very **little visible change** at the systems-level but a **great deal of experimentation** at the individual level; a take-off phase where the process of change starts to build up and the state of the **system begins to shift** because of different reinforcing innovations or surprises; an *acceleration phase* in which **structural changes** occur in a visible way through an accumulation and implementation of socio-cultural, economic, ecological and institutional changes; and a *stabilization phase* where the **speed of societal change decreases** and a new dynamic equilibrium is reached."[12][7] Against this background this paper will describe the starting point for formulating long-term, collective innovation strategies in the EIIS, the actors involved, the exploitation of vision formulation, the expectation for visions to become leading images for corporate orientation, the phases of transition which have already been initiated on the way to a sustainable electronics industry, and the main driversŽ contribution in this process.

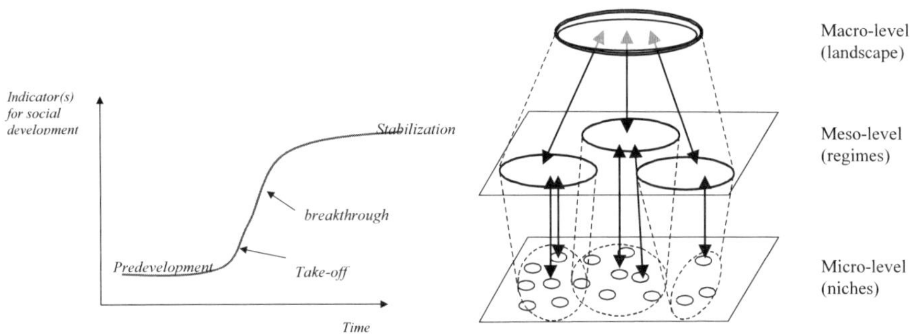

Fig. 3. Four phases and different levels of transition (Geels and Kemp 2000)

(2) Transition takes place at different levels, influencing each other: the micro, meso and macro level (see again Fig. 2).[5] The micro level (*niches*) relates to individual actors, companies, and technologies, referring to the place where novelties are invented, tested, and exploited. The meso level (*regimes*) relates to networks, communities and organisations, institutional arrangements, dominant practises, rules, and shared assumptions.

[7] In Chapter 3 empirical evidence is given for drivers and reaction patterns of the innovation actors in the EIIS (text highlighted by the author).

At this level, also technology, production, user, and policy regimes are distinguished [2]. The macro level (*socio-technical landscape*) comprises conglomerates of institutions and organisations (e.g. a nation) and relates to material and immaterial elements like infrastructure, political culture and coalitions, social values, macro economy, demography, and the natural environment.[8] In the EIIS, transition takes place at all these different levels, and this paper will sketch some of the ongoing developments in – an overall view – account for the 'system innovation': at the micro level, changes in the corporate innovation strategies take place, including the increasing relevance of sustainability aspects in regular innovation management procedures. At the meso level a dynamic interplay of institutional and technological change takes place, modifying the entire system of waste and recycling management and, at the same time, changing the market and actors configuration substantially. At the macro level the expectations and requirements of society regarding sustainable development are an important driver. Factors like demographical change, global warming, etc. result in additional challenges the EIIS must face and are expected to provide solutions for (e.g. electronic devices suitable for an aging population and minimising energy consumption).

(3) Usually, "transition denotes a long term change process in an important subsystem encompassing various functional systems (e.g. food production and consumption, mobility, energy supply and use) in which both the technical and the social/cultural dimensions of such systems change drastically." [11] In this paper, the EIIS as the subsystem is tackled, embracing various subsystems (technology, governance, supply chain, market, etc.) asking for the drivers of the long-term change and their respective contributions to move the EIIS towards sustainability.

(4) Since the Dutch experiences have evolved through implementing transition management only on national level and in one important governance regime (4th National Environmental Plan), it does not mean that problems of transition and transition management are restricted to a national scope. On the contrary, transition processes often appear on a global scale, within internationalisation and globalisation processes and as the result of global environmental problems disregarding national borders. So the question of transition management in the EIIS is not restricted to a particular national scope since the transition is expected to take place in a world-wide context. Following the international nature of transition this paper will therefore ask for the drivers for transition, and in this respect, will draw on the EU environmental policy.

(5) The transition research holds out the prospect of analysing socio-technical changes in a more interrelated way than a variety of mono-disciplinary approaches.[9] Indeed, to explain system innovations, a more interdisciplinary

[8] See [2], p. 005, also [13]

[9] See [11], p. 1.

approach is needed to investigate the impact of technology development, infrastructural changes, modifications in market transaction structures and actors relations, alterations in behavioural patterns, in cultural values etc. In the EIIS, the most promising pathways to sustainability require collective transformation such as new product-service systems (PSS) connected with new business models terminating the present value-added chain and introducing new co-ordination structures beyond market transactions and hierarchy.

(6) "Transitions cannot be managed in the strict sense, i.e. they cannot be steered by a central actor (government or other) to realise specific objectives ... By implication, transition management is an interactive process that needs to take place between heterogeneous set of actors, each acting on the basis of their own vital interests and expectations".[11] Using the example of EU environmental policy transfer to national member states in the case of the WEEE/RoHS[10] for the EIIS, this paper will also give some empirical examples of how the process of political decision-making interplays with specific actor relations in the EIIS, how the process of developing particular (new) formal governance rules and institutional arrangements is to be characterized in terms of actors‘ cooperation, how the governance regime interplays with corporate decisions and technology regimes, and how – by the end of the day – transition of the system evolves in the face of new institutional set-ups.

4 Transition Processes in the Electronics Industry Innovation System

4.1 Breakdown of Transition Management for the Electronics Industry

The concept of "transition management" is defined as = current policies + long-term vision + vertical and horizontal coordination of policies + portfolio management + process management [2]. Transition management thus is characterised by the following issues:[11]

- Evolutionary steering concept (governance, interactive government, networking)
- Multi-actor governance (aims at system innovation and sustainability)
- Adaptive and anticipative management (uncertainty and complexity management)
- Steering through learning (doing-by-learning and learning-by-doing)

[10] RoHS = (EC directive on) Restriction of Hazardous Substances, which is an "add-on" to the WEEE directive restricting the use of certain hazardous substances in electronic products or parts like mercury, lead etc.

[11] See [2] and [14]

- Orientation towards transition goals (less short-sightedness)
- Orientation towards learning and innovation (to overcome the preference for quick results and policy reliance on technical deadlock)
- Alignment of different policy domains to overcome fragmentation
- Programmes for system innovation based on visions of sustainability
- Opening up for policy process (to decrease domination by vested interests)

The transposition of the "new EU approach in waste legislation" into the member states and the implementation of take-back ordinances and other policy instruments to change the innovation system according to resource consumption, material streams, and substance flows may be addressed as an example of transition management, since it embraces more or less all of these elements in a long-term and multi-level governance approach. The focus of this paper lies on institutional change in the transition of waste management in the European Union, using the example of introducing take-back obligations in the Electronics Industry Innovation System. Special attention is devoted to the interplay of:

Changing current waste policy concepts from linear streams to circular loops

+

Defining long-term goals within the process of implementation of new policy instruments for recovery, resources consumption, eco-design for products, etc.

+

Vertical and horizontal coordination of policies by combining the WEEE-directive vertically with the RoHS-directive within the implementation process, and by embedding the WEEE into other horizontal environmental policies like IPP (Integrated Product Policy)

+

Portfolio Management using institutional change defined as setting up particular negative rules as well as a set of permitted ones [15] (as laid out in the WEEE directive) to avoid cognitive lock-ins and make use of markets for coordination and context control instead of planning.

+

Process Management with government adopting different roles in the transition phase.

These aspects of transition will be tackled within the next chapters.

4.2 The EU Sustainability Policy with Regard to the EIIS: Development of the EIIS Governance Portfolio

System innovations involve problem areas of competition, environmental, and employment issues, thus cross-cutting different governance arenas and overtaxing political decision making [16]. The governance shifts that took place during the last two decades in the EIIS went along with increasing complexity in the innovation system. It turns out that the political process as well as the effective implementation of single regulations[12] in view of the sustainability vision is an extremely complex multi-level task that needs participatory elements since otherwise the political process would be indeed overtaxed.

First of all, it should be stressed that in the transition process of the EIIS, there is no single or collective transition manager.[13] Since we are facing a multi-level governance (EU, national level, regional level) and multi-level innovation systems (global, EU-wide, national, regional), the transition management appears to be 'virtual' in a sense that political decisions are developed using multiple communication channels, consulting arenas and lobbying structures on all governance levels mentioned. Facing the scope of the transition task it seems obvious anyhow that a single, or even collective transition manager may be overtaxed with the 'steering' of the EIIS transition.

How was the governance arena set up to direct the innovation system towards the new paradigm of a circular and sustainable electronics industry? What is the composition of the governance portfolio and how does it interplay with technology, market and societal drivers? In Fig. 4 the most relevant innovation drivers of the EIIS are simultaneously specified:

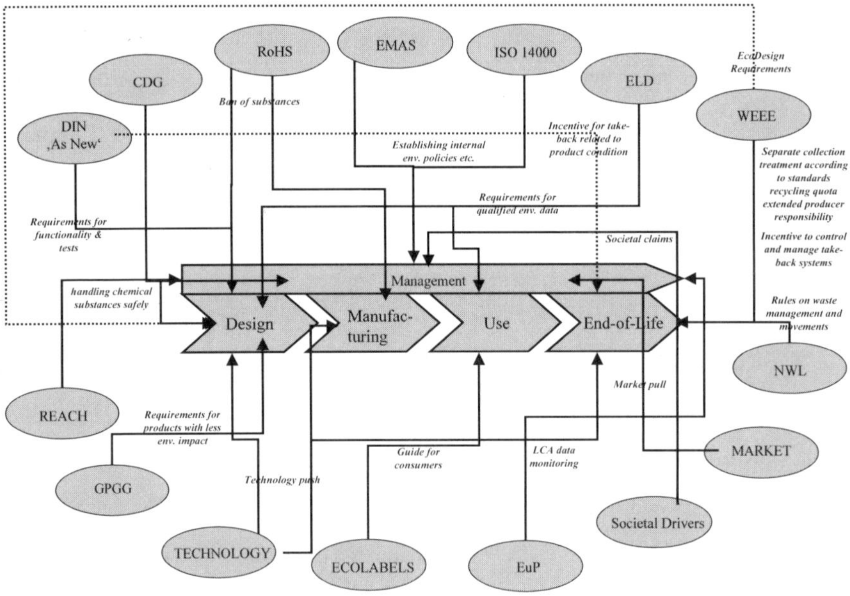

Fig. 4. Innovation drivers in the Electronics Industry innovation system

[12] In this paper, I will only draw on the WEEE and the RoHS as an example for a multi-level implementation of EU directives concerning the European, as well as the national and regional level within the member states.

[13] As mentioned in the Dutch reports, see [2]

The WEEE-Directive take-back regulation influences different stakeholders of the innovation system by introducing direct legal obligations to be fulfilled, such as collecting and recycling *quotas*, the implementation of the '*producer responsibility principle*' and *financial responsibility* for take-back systems, the definition of certain *standards* for waste management, as well as several requirements concerning *labelling* of products and data, and mass flow *monitoring*. Manufacturers of electrical and electronic equipment are burdened with the costs of collecting their end-of-life equipment leading to considerable pressure to re-structure the product design through easy and low-cost disassembly, the end-of-life (EOL)management by establishing new logistical concepts, take-back systems and recycling systems, the innovation management by introducing new environmentally oriented requirements like Design For Environment (DFE) within the supply chain, etc.

The RoHS directive operates with *prohibitions* and restricts the use of certain hazardous substances in electrical and electronic equipment, as for instance lead, mercury, and other heavy metals with a considerable impact on the manufacturing process and recycling requirements.

The EuP directive places a strong burden on companies which produce energy-intensive products to meet *environmental requirements* and targets in the product's design, production, and end-of-life phase. The EuP requires an assessment of the ecological equipment profile (LCA) regarding raw material, acquisition, manufacturing, packaging, transport, distribution, installation, maintenance, use and end-of-life treatment.

The upcoming chemical regulation REACH (**R**egistration, **E**valuation, **A**uthorisation and Restriction of **Ch**emicals of the EU) asks for a registration of all relevant chemical substances in the supply chain. Manufacturers and importers have to demonstrate in a registration dossier that they manage their chemical substances safely.

GPGG (Green Purchasing Guidelines of Governments) in place are still increasing the demand for products and services with a lower overall environmental impact. These guidelines put pressure on producers to develop products and services causing less environmental damage.

The CGD (Customer Guarantee Directive) requires that consumer goods conform with the contract of sale, that they are repaired, replaced, or a refund is given if a defect becomes apparent within two years of delivery, and that contractual guarantees comply with certain criteria. This places a heavy burden on design for quality and functionality.

The DIN "As New" standardisation is still in progress: It concerns the dependability and quality of products containing re-used parts and places additional requirements on functionality and tests.

The focus of ISO 14000f is on establishing internal policies, procedures, objectives, and targets, and on pursuing continual improvement. EMAS goes beyond ISO 14000: Organisations registering to EMAS must be able to demonstrate that they have identified and know the respective implications of all relevant environmental legislation and that their system is capable of meeting them continously.

There are three types of Eco-labels. Type I is a guide for consumers identifying products as being less harmful to the environment compared to others of the same function (i.e. German Blue Angel, Nordic White Swan etc.). Type II sets up requirements for self-declared environmental claims including statements, symbols, and graphics on products or services, which are not certified by an independent third party (i.e. "recyclable", "biodegradable" as examples of statements; "Mobius Loop" as an example of a symbol). Type III requires a set of quantified environmental data consisting of pre-set categories of parameters based on life cycle assessment according to ISO 14040.

The ELD (Energy Labelling Directive) requires that appliances shall be labelled to show their power consumption in such a manner that it is possible to compare them with other brands and models in term of efficiency with that of other makes and models (appliances for household use).

Electrical and electronic equipment will be affected by Integrated Product Policy (IPP) as well. It represents a new approach for product-related environmental policy and advocates life-cycle thinking which means that consideration is given to the whole of a product's life cycle from cradle to grave. IPP seeks to minimise environmental degradation by looking at all phases of a productsŽ life cycle and taking action when it is most effective (design, manufacturing, use, disposal).

Besides *market drivers* there is also a series of *societal drivers* which increase the awareness of environmental problems among producers and consumers, caused by the public debates following environmental accidents (e.g. Three Mile Island and Chernobyl, Exxon Valdez, Brent Spar), a fear of scandals (like the problem of transborder waste shipments to Eastern Asia) and 'stakeholder claims' (such as local, national or international NGO's activities).

With respect to system innovations it is important to recognise that regulation, interplaying with the other drivers, can both facilitate or hinder the incentive features of both technology push and market pull effects [17]. For instance, in the EIIS innovation effects are increased substantially by combining several policy instruments directed towards the early design phase of product innovation, demanding on alteration of product functions to decrease energy consumption during the use phase as well as disassembly and recycling costs at the end-of-life phase. So, especially the WEEE, the RoHS and the EuP as regulative drivers have a strong impact on the transition of the EIIS from a linear to a circular economy providing striking incentives for new business models, recycling, and re-use as elements of the system innovation.

4.3 Transition Phases in View of the EIIS Governance Portfolio

The governance portfolio in Fig. 4, as it determines the EIIS today, developed over time as the result of increasing regulation intensity in the 80ies and especially in the 90ies of the last century. The transition phases – covering the WEEE, the RoHS and the EuP directives – are depicted in Fig. 5.

Pre-Development Phase – Setting the Scene

In most of the OECD countries, environmental policy measures were initiated in the period between 1960 and 1970. As environmental protection has occured as an independent policy regime in several industrial countries since the beginning of the 70ies it has been denoted as "environmental policy" since 1970 [18]. The first set of policies, their style, the instruments used, as well as the responses from the innovation actors were primarily "reactive" and directed towards establishing an information and consultancy infrastructure, e.g. in the area of waste management. A vast majority of prohibitive regulation connected with a command-and-control policy style can be found up to the late 80ies and even in the 90ies. In response to the oil crisis in the 80ies predominantly process improvement and resource optimisation as "receptive" behaviour were pushed, involving managers as the main innovation actors. In the EIIS this institutional framework of reactive policy style operating with prohibitions, limit values, etc., together with incentives from public research funding (in EU programs, national technical funding in Germany) led primarily to incremental innovations in end-of-pipe technologies within the area of material and substance recovery in end-of-life management of electronic products [19]. Main activities comprised technical solutions from the range of the sorting, conditioning and recycling techniques for metal waste, process technique for the recycling of plastic wastes, processing of galvanic baths and etching solutions as well as thermal waste treatment.

Take-Off Phase – The System Begins to Shift

Since the beginning of the 90ies and with the publication of the Brundlandt report in 1987, a change in the policy style has appeared, referred to as "ecological change", or "ecological modernisation" supported by a comprehensive societal perception and public discourse reflected in the media. For the EIIS at that time the crucial innovation driver was the *policy style* at the end of the 1980ies: the German Packaging Ordinance under the old Waste Management Act had already mobilised a great deal of changes in the materials flow for (used) electrical goods. The presentation of the first draft "Electrical Scrap Ordinance" in Germany in 1991 (see Fig. 5)[14], embodied in the Closed Substance Cycle Act and the implementation of the Packaging Ordinance also gave political weight to the prospect of a rapid application of material flow regulations for the electronics industry. This gave the industry clear incentives to organise itself.

[14] Please compare with Fig. 6 and Fig. 3.

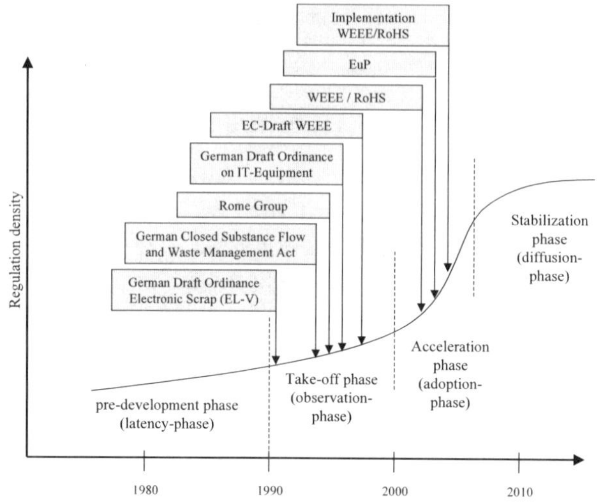

Fig. 5. Transition phases of the EIIS according to the governance portfolio of WEEE/RoHS and EuP

Following a "constructive" reflex the industry developed new approaches of transforming environmental concerns into opportunities for selling new products and services. Green marketing appeared, a considerable amount of green products were launched, especially in the white goods sector (example: energy star refrigerators) of the EIIS. In those days, "attempts to improve the environmental performance of technologies tended to emphasise processes of innovation associated with individual technologies. The focus tended to be on switches from more polluting to less polluting processes and products." [6] Nevertheless, these innovations are to be characterised as incremental, since the technical solutions derived from these activities were not radical in the sense that they describe and follow a different trajectory.

Starting from 1994 the innovation activities diversified increasingly and led to more extensive project initiatives strengthened in the front-end range. Various improvements belonged to them, such as in particular recycling and disassembly-friendly development and construction, the development of tools and analysis instruments (software for DFE (Design For Environment), economic analyses of the disassembly depth, analyses of the recycling ability, etc.); projects for partial automation (disassembly, dismantling) and improved partial recognition, systems for secondary raw material use, DFE and recycling, selected individual products with high mass accumulation (greener television etc.), single techniques for improvement of the separation. [16] Starting from 1995, a certain concentration of problems in the material flow management showed up and resulted in an intensified effort regarding front-end solutions

[16] Adapted from [12].

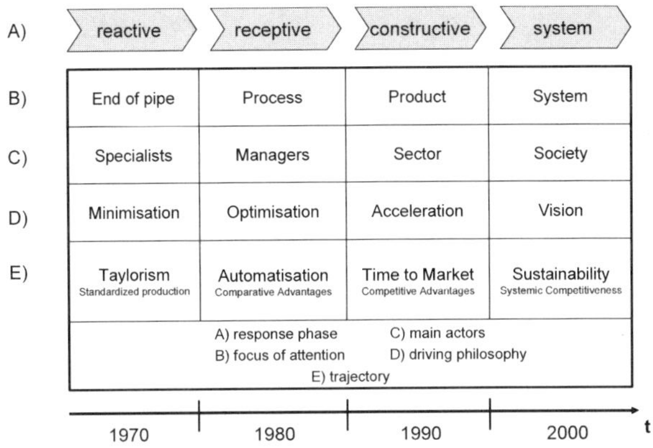

Fig. 6. Development stages in corporate and societal response[15]

within the range of construction and production, connected with a transition to questions of upgrading, re-use, use intensification and life span extension of products. These innovations may be addressed as "radical innovations" since the concepts of re-use and life span extension are superceding the "time-to-market" and "short innovation cycles" trajectory (first sign of a paradigm shift indication towards circle economy).

Acceleration Phase – Visible Changes in the Innovation System Take Place

In the years 2002-2004 visible structural changes took place in the EIIS. In most of the former member states of the EU (before May 1, 2004), take-back measures according to the WEEE have been implemented, the transposition of the WEEE directive into national legislation has been in progress, manufacturers are now engaged in DFE programmes, market green products, develop supply chain measures, and so on.

Starting with the new millennium, an outline of a new innovation paradigm turned up in the EIIS: an orientation towards global sustainable development requiring "system innovations" in the EIIS. Since then a global strategic re-orientation has taken place affiliated with new requirements throughout the whole supply chain, placing new challenges on networking with new business actors (like waste-management companies, recycling industry), changing manufacturing processes and product design (e.g. for energy efficiency), "thinking green" and changing innovation management procedures, developing new business models for a product-service shift and so on. These changes are described in more detail in Chapter 5, as they refer to changes in market and actors configuration (Chapter 5.1), changes in technological regimes (Chapter 5.2) and in behavioural regime transition (see Chapter 5.3) on the meso level.

Stabilization Phase – Towards a New Stable Situation with a New Equilibrium?

The stabilisation phase of the EIIS transition from a linear to a circular economy is supposed to arise after the implementation of the WEEE/RoHS and the EuP on the national level. This process is expected to take again a decade, since the WEEE and the RoHS have individual time schedules for monitoring and adjusting the set according to technological progress etc. However, it is also expected that the governance portfolio as described in Chapter 4.2 will provide sustainable dynamic incentives for the innovation players beyond compliance issues. Insofar, the EIIS will progress within the guard rail of new business models putting incremental innovation into practice.

5 Meso-Level Transition: Pathways to Sustainable System Innovations in the Electronics Industry

5.1 Shift in Market and Actors Configuration

To clarify the shift towards system innovations, a review of the development of the actor configuration in the EIIS is important. Parallel to the upcoming new innovation drivers and their cognitive processing, the innovation system became substantially more complex in the transition from the last decade of the 20th to the first decade of the 21st century based on the structure of the actor configuration. The typical actor configuration up to the end of the 90ies represented a linear progression of different added value and EOL processes.[17] The substantial innovation drivers in this innovation system have been economic incentives such as "market pull" (demand changes, price signals, etc.), as well as "technology push" on the part of research (e.g. microelectronics), which jointly stimulate the development of incremental innovations such as new products and procedures.

In the transition to the first decade of the 21st century the innovation system exhibits a strongly extended actor configuration, due to the development towards radical innovations and system innovations activated through the regulation context leading to a cycle-structured economy: starting from an original linear supply and value added chain and the associated actors, new actors from the range of waste management, recycling economy, as well as service in the re-use area step in.

As displayed in Fig. 7, after the millennium change the EIIS presents itself as a complex network of actors and institutions. Those actors are at first the large manufacturers of electronic devices (such as Sony, Philips, Sharp, Miele,

[17] The grey boxes in Fig. 7 on page 72 represent the elements of the "old actors system" before the millennium change, consisting of a linear supply chain from raw material suppliers via producers, assemblers, users to landfill ("throw away" economy).

etc.), their suppliers in the supply chain, e.g. components manufacturers (like Infineon, ECM, AMD, Bosch, Intel), second tier suppliers, like the chemical industry or subassembly manufacturers, research and development institutes, technology transfer companies, consultants, banks, insurance companies, recycling and re-use companies, maintenance and repairing service providers, logistics companies, manufacturing devices producers, waste processing companies, and so on.

For all actors mentioned there are interconnections like economic transactions in the relationship between manufacturers and customers, as well as institutional arrangements to coordinate the innovation process (professional organisations like the EECA (European Electronic Components Manufacturers), BITKOM, ZVEI or R&D networks (ECOLIFE-thematic network). The innovation process is influenced by all these actors. The examples given later demonstrate the systemic character of the innovation process in the electronics industry.

With respect to system innovations, it can be derived from Fig. 7 that the knowledge genesis and its conversion for triggering innovations take place in a complex network of various actors, who join their different core capabilities in the innovation process. Besides the complexity of the actor configuration, the complexity of the incentive structure and the driver for innovations rise as well. Due to anchoring the EPR principle (extended producers' responsibility) in the minds of the manufacturers, the innovation actors involved are globally directed to a stronger environment and sustainability orientation. As a central actor of the innovation system the manufacturer undoubtedly determines the direction of the innovation.

The transition to the new actor configuration developed gradually, beginning around the early 90ies as a reaction to the announcement of environmental political regulations within the range of the closed-loop recycling management, and is not yet terminated. To that extent, Fig. 7 rather represents a snapshot. In particular, the development of new business models starting at the end of the 90ies will further merge more actors in the innovation system, who, at present, either do not play any role in the EIIS, or only a minor one (e.g. 'content provider' in the ICT sector).

5.2 Technological Regime Transition

Looking back at the last 2-3 decades, the technological regime transition in the EIIS describes a change from a more or less stable socio-technical configuration to another, to be characterised by a shift from functional to a systemic product-service economy. Before the millennium change the EIIS was devoted to a stable functional trajectory, based on the ICT paradigm up to the late 90ies. Relying on a huge preceding technology push provided by the dramatic progress in microelectronics, it developed as a result of constant incremental innovations such as step-by-step improvements of technology (i.e. improvements of the technical efficiency of parts) and products (improvements of

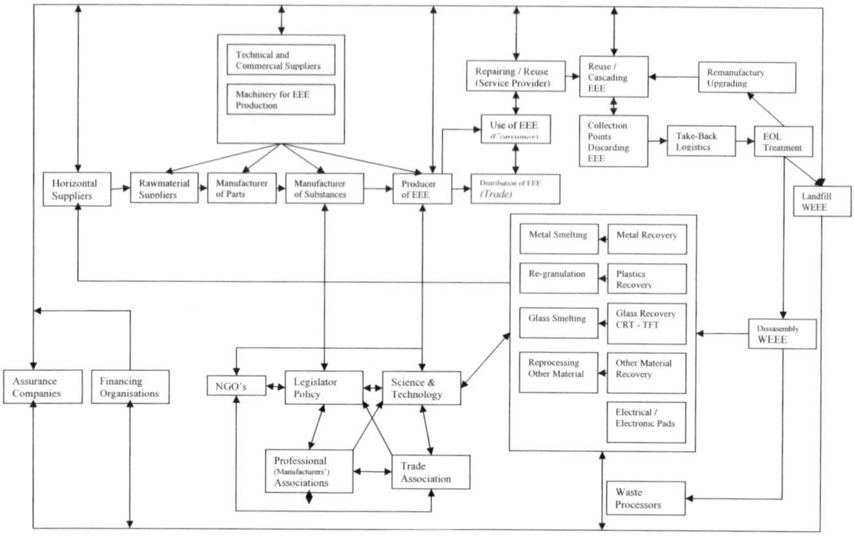

Fig. 7. Circular innovation system in the electronics industry after the millennium change

functions to better fulfil consumer needs, i.e. white goods), and – driven by increasing demand – framed a vast time-to-market innovation regime combined with a linear 'throw-away' economy.[18]

Radical technological innovations (e.g. analogous to digital shift, broadband communication, etc.) have provided the set for changing the trajectory in the last five years leading to a great choice of possibilities in the layout of new business models (technology push). Insofar the digital revolution forms a new paradigm with service extensions (GPS, interactive TV, etc.) which is expected to provide persuasive advantages in terms of sustainability.

The 'seeds for transition' from a functional production- to a service-oriented and knowledge-based EIIS are to be found at the:

- micro level of companies, as innovators from different sectors of the economy (like communication providers, content providers, waste management, recycling etc.) – by gaining for example from the fusion of different ICT media (audio-video communication etc.) – push into the EIIS, fostering impulses and incentives for core innovation actors (EEE manufacturers);
- meso level of networks, as innovation actors emphasise joint cooperative innovation processes and intensified pre-competitive collaborative research and development programmes, with horizontal and vertical players in the supply chain, as well as with scientific actors beyond technical research.

[18] See again Fig. 6.

Besides technological shifts, enabling product-service systems, also new technological know-how is needed in the EIIS. Due to the implementation of the new governance portfolio (WEEE, RoHS, and EuP), a shift in the fundamental set-up of the value-added chain was introduced, moving the EIIS from a linear to a circular value-added chain and unfolding the need for additional service systems to close the loops at different stages of an electronic or electrical product, i.e. in the use phase by setting up new maintenance, upgrading and re-use services, as well as closing the material circle by introducing take-back and waste management services.[19] This sets the scene for introducing new recycling technologies and take-back systems (according to the WEEE) as well as new manufacturing processes according to the phase-out of certain hazardous substances to comply with the legislative requirements (RoHS). It demands also the involvement of new core competences and knowledge from new innovation players in the EIIS. Here, the core competences of the manufacturer are not sufficient anymore. On the way to a circle economy and for a "design for recycling", detailed information for the disassembly processes, for the separation, and for the disposition of materials in end-of-life treatment etc. is needed. To that extent, the knowledge of recycling enterprises about the innovation process becomes indispensable.

Which are the "new technologies" and innovation topics tackled actually by the innovation actors? Table 1 depicts a selection of 42 out of 120 innovations, rated as most important for the electronics industry as the result of an experts delphi within the EU-ECOLIFE II network. These innovations are embedded into the main stages of the life cycle of an electronic product, i.e. 'Design', 'Manufacturing', 'Use', 'End-of-Life', and 'Management', defined as a horizontal task throughout the life cycle. At the same time, their sustainability potential is rated, using selected economic, ecological, and social indicators[20]:

Some of the innovations mentioned in Table 1 are narrow in scale and scope (like a particular plastic separation technology), others are complex because, as a pre-condition for their diffusion, an extensive alteration of business structures, management procedures, and infrastructural conditions has to be worked out in the innovation system (for example: new use concepts and product-service shifts require a complex cooperation of various innovation actors along the supply-chain involving additional players from service and maintenance). Insofar, the whole range of different innovation types is covered, such as technological innovation (e.g. further development of a certain recycling technology, EcoDesign-tools), organisational innovation (new logistics concepts for take-back systems, new green management tools), personnel innovation (i.e. new education and learning concepts), and systemic institutional innovation (new business models).

---

[19] See again Fig. 6.

[20] as a result of an ex-ante evaluation in the ECOLIFE II network

To sum up, we learn from the scope of the "innovation and technology landscape" depicted in Table 1 and from the shift in market/actors configuration[21] that, in the technological regime transition, the EIIS proves to be receptive to involve new players and their technological know-how and that – how the work on the sociology of technologies puts it[22] – suitable [24] tech-

Legend: expected strong impact; expected medium impact; expected low impact; expected indifferent

	Innovation topics in the Electronics Industry Innovation System	Sustainability Indicators: Econom. – Efficiency	Productivity	Transaction Costs	...	Ecological – De-Materialization	De-Toxification	De-Energetization	...	Social – Health & Safety	Working Conditions	Encouraging Learning	...
Design	Ecological idea dissemination through the supply chain												
Design	Eco-Co-Design with Suppliers												
Design	Management of Eco-Cost Reduction with Suppliers in manufacturing & design												
Design	Communication strategies among companies												
Design	Information dissemination to SME												
Design	Design for Environment												
Design	Design for EOL, dis/assembly												
Design	Integration of DFE in conventional management systems												
Design	Substitution of hazardous materials (e.g. BFR, VOC's, semi-conductors)												
Design	Renewable materials												
Design	LCA/LCC including simplified LCA												
Design	Database on Materials/Components for DFE												
Design	Life Cycle Engineering												
Design	New Substrates for PWB												
Design	Halogen-free flame retardants												
Design	Mercury Free Light for Flat Panel Monitors												
Design	New Flame retardands materials												
Manufacturing	Dissemination of best industrial process												
Manufacturing	Substitution of hazardous materials												
Manufacturing	IPPC												
Manufacturing	Improved Manufacturing of materials, components & subassemblies												
Manufacturing	Lead-free soldering												
Manufacturing	Eco-Efficiency of Manufacturing												
Use	Customer Information and Education on usage												
Use	Communication of products impacts to the consumer												
Use	Understanding Customer Behaviour And Communication with Customers												
Use	Energy Efficiency in Use												
Use	New Business Modells (Leasing etc.)												
Recovery / EOL	Information communication between Electronics Industry and Recyclers												
Recovery / EOL	(Cost Effective) EOL and Recycling Technologies												
Recovery / EOL	Standards and Technical Specifications for Recycling												
Recovery / EOL	Logistical concepts concerning collection of used electronics												
Recovery / EOL	Disassembly Analysis												
Recovery / EOL	Recycling of materials and components, special interest materials												
Recovery / EOL	Development of (public) take-back schemes for EOL												
Management	Supply Chain Management												
Management	Knowledge Management, Knowledge Transfer and distribution												
Management	Education and Training												
Management	Legislation monitoring of RoHS, WEEE, IPP, EEE etc.												
Management	Ensuring legal Compliance												
Management	Green Strategy making and Green Innovation Management												
Management	Ensuring legal Compliance of Suppliers												

Table 1. Innovations in the Electronics Industry innovation system rated as most important[23]

[21] See again Chapter 5.1

[22] A comprehensive overview on this issue is given by [20].

nological choice is made within 'systems' or 'networks' that "involve not only the firms which manufacture the technological products themselves, nor just their suppliers and large customers, but an extended and heterogeneous array of investors, regulators, unions, professional associations, government departments, research, and educational and political organisations."[25]

The main drivers of the technological transition arise from within the regime itself and from the legislative context. At first glance, the transition towards a service-oriented EIIS seems to be technology-driven in addition as will be demonstrated, there are other regime shifts necessary to move the EIIS from a hitherto stable configuration to another.

5.3 Behavioural Regime Transition

The 'value and belief regime' of an innovation system is usually characterised by a set of formal and informal rules (e.g. business routines, tacit habits), cognitive frames of the innovation actors (e.g. agreed common understanding of the way of life), innovation and design ethics as shaped from education and over time, behavioural attitudes of customers moulded by rules of using technologies, products, and services, as steered by intrinsic and extrinsic incentives with respect to the evolvement of needs, etc.

These beliefs give shape to the function and purpose of technologies, since "technologies have a purpose (a function) that is not instrinsic to their physics, because physics has no conception of purpose, it just 'is"[22]. This shaping of technologies, in their final artefact outcome as manufacturing processes, products or services, undergoes an interaction of expressions of needs from the customer side and tacit knowledge of the designers in innovation processes. At first glance and to put it crudely, the evolvement of innovations is both 'demand-driven' and relies on fundamental design concepts, such as the 'operational principle' (function that the device must fulfil) and the 'normal configuration of the device' (general shape and arrangement which are commonly agreed to best embody the operational principle).[26]

We will touch two major shifts in belief regimes interplaying with the described technological regime shifts and the shifts in the governance arena: (1) change in design trajectories and (2) change in business trajectories.

Change in Design Trajectories

Following the innovation paradigms of former waves of development[27], it is obviously the design tradition – as a result of a stable socio-technical configuration in the last 2-3 decades – that led to the functional trajectory of the

[24] This table comprises the result of a technology experts delphi (32 experts), conducted in 2003 in the ECOLIFE thematic network (see for details [21]).

[25] [21], p. 30.

[26] [22], p. 3.

[27] See again Fig. 6.

EIIS as described above and to a cognitive lock-in from the design side. At the same time, consumer needs are changing rapidly at first glance (fulfilment of new needs like mobile communication, remote access to personal data) but their habits became fixed in longer waves of development ("ownership is better than occupancy"). So, it is explicable that fundamental changes in satisfying consumer needs did not appear on the agenda unless customers explicitly demanded new ways of satisfaction. This happened for example early in the B2B area regarding leasing concepts of computers and result-oriented services, for instance, in the machine building industry in the provision of operating supplies.

The shift in belief regimes obviously needs more time to take place than the technological regime shift. At the same time, the belief shift requires additional incentives from the socio-technical landscape, i.e. from the macro level of society (development of a new common understanding) and must be supported by other meso level policy measures (e.g. education, R&D programmes etc.). This is actually an ongoing process, for instance, in Germany and at the European level, as targeted research and development programmes are implemented to promote the search for socio-ecological solutions and new sustainable business models.[28] Transition management has to modulate these dynamics in an innovation system in the same normative direction. The laundry example given later in Chapter 5.4 addresses opportunities for the industry to support belief shifts by setting up appropriate communication strategies. This (by the way) happens at present, initiated also by the new governance context in the EIIS[29].

Change in Economic and Business Trajectories

Another major belief shift is required when looking at new business models like Product-Service Systems (PSS). The evolution and diffusion of these new sustainable business models depend on radical changes in economic paradigms and requires a change in the perception and in the behaviour of all actors involved in the innovation system.

What is this "new economic paradigm"? For the dissemination of system innovations as referred to as new business models in the EIIS (e.g. life-cycle extension, durable products, product-service shifts etc.), new incentive systems have to be implemented to shift earning possibilities from "old economy strategies" (earnings as a result of shortening the innovations cycle) to "new sustainable economy strategies" (earnings as a result of life time extension, energy minimization, intelligent services etc.). The value added in new sustainable economy strategies is located in new service systems providing life-time

[28] As an example the new R&D programme "PRONA" launched by the German BMBF in 2004 is expected to cover a great deal of the overriding questions concerning progress in the long-term vision for sustainable development.

[29] As the result of the Eco-Labelling directive, see Chapter 4.2.

extension via repairing, maintenance, service for energy minimisation, and so on. Furthermore, value added may be created by multi-generation product planning, time-dependent product innovation systems with cascades of product use, re-manufacturing, and refurbishment options. The consequences for production, distribution and marketing are tremendous. The whole innovation process from R&D to distribution and sales demands to be revised and needs a completely new actors configuration as well as joint development activities between these new actors.[30]

The success factor "time to market" is strongly connected with the old paradigm where innovation cycles are short and pressure for new product launches is high. To gain early-bird advantages and first-mover profits it is indispensable to push technology, to shorten R&D cycles, and to realise a fast product launch. The whole innovation system is adjusted to this economic paradigm, a paradigm that equals earnings with throw-away behaviour. In the new economic paradigm, the innovation system will be detached from "time-to-market" as the key issue of economic strategy. That does not mean that there are no first-mover advantages to beat the competitor. There is just another strategic orientation of the business model: it is set up to earn money with intelligent services around a product over its life cycle.

It requires however an alteration of

- R&D, which must be adjusted to longer life cycles.[31]
- Production, which must, for instance, be adjusted to multi-module products to be assembled for customisation.
- Distribution and marketing, which must be adjusted to selling services and further benefits for the customer (functions) instead of selling a product or technology.

Bearing this in mind, it is obvious that, within system innovations like the evolvement and diffusion of new business models or PSS, a co-evolution of technological, institutional, behavioural, and cultural regimes has to take place, since, in particular, new use systems need behavioural changes on the side of the customer. Also – coming back to the differentiation of the micro, meso and macro level of transition – it appears to be quite granted that the belief regimes on the meso level are expected to gain superior attention, since normative rules and shared assumptions are the main important drivers for new business models on the demand side.

[30] The Gotland experiment of Electrolux described later in Chapter 5.4 gives at least some ideas on how these business issues have to be changed to successfully implement new business models.

[31] Compare the paper on deceleration by E. Günther and M. Lehmann-Waffenschmidt in this volume.

5.4 Product-Service Systems (PSS)

As potential benefits of new business models like PSS the following are usually mentioned:[32]

- Easier access to product function via "pay per use" without investment,
- More use with fewer resources by reduced downtime due to collective use (sharing, pooling),
- Extended life cycle due to repair maintenance, integration of reused components or recycled material,
- Improvement in customer relationship management,
- New business opportunities and new ways of profit generation.

Examples of more advanced product-service systems[33] are shown in Table 5.4[34], a more detailed example of the co-evolution process of niches, regimes etc. using the example of laundry [23] and the Gotland experiment of Electrolux[35] is described afterwards:

As shown in Table 1, the sustainability potential of new business models (like for instance PSS[36]) in the EIIS is assumed to be superior, though rebound effects may also appear in setting up new product-service systems when the demand for certain products and services is increased occasionally. In so far the empirical ex-ante evaluation of expected sustainability effects of single innovations needs further empirical tests and validation. However, beyond compliance measures for existing regulation and as the result of the new governance portfolio in the EIIS, new business models are presently the main topic on the agenda of nearly all major EEE manufacturers.[37]

In the Gotland experiment[38] Electrolux started a field trial on functional sales "pay per wash" in November 1999 to December 2000 to evaluate if households are interested in paying for the service instead of purchasing the appliance itself. The model included a washing machine delivered to the consumer and a "pay per use" fee of 1€per wash at 1 kWh/wash cycle. The consumer paid for the installation itself a cost of 45€. Electrolux offered an additional service for repair and maintenance (24 h) as well as a new washing machine after 1000 wash cycles. The business model also involved a refurbishing of the used machines at the Electrolux refurbishment facilities. The energy supplier

---

[32] Expert interview with Motorola within the EU-ECOLIFE project.

[33] That means, product-service systems which go beyond product-oriented services, e.g. service integration, which is already standard in industry.

[34] EU-ECOLIFE project, samples provided by SAT, Vienna.

[35] Electrolux has conducted a field experiment in Gotland.

[36] See for a description of PSS also [24], p. 4 ff.

[37] One of the main activities in the EU-ECOLIFE thematic network is the evaluation of product-service systems and new business models. In this investigation – besides research institutions – all global players like Motorola, Sony, Continental, Phillips, Electrolux, Schneider, Fujitsu, Merloni etc. are involved.

[38] See details in [25]

Players involved	New Business Model / Product – Service System
Solvay Pharmaceuticals, Dell	Dell provides pre-configured notebooks to the spread workers of Solvay. Solvay leases the equipment and upgrades it at end of the leasing contract.
Xerox	Xerox offers companies to lease office equipment (copy machines, printers, etc.) with option to upgrade. Xerox also has a sophisticated end-of-life product management.
AB Electrolux (Electrolux Euroclean)	Electrolux provides household appliances (mainly cleaning devices) for rental to business and private customers. It also puts focus on product optimisation (Gotland experiment)
IBM	Leases IT equipment, mostly for business customers. At the end of lease several options can be chosen: upgrading, extending the lease, purchasing the equipment, or returning it. The equipment is mostly returned and then remarketed.
Alfaskop	Application service provider. The company provides computer power and applications over the Internet – the product is the constant computer accessibility.
Thorn (retailor)	Lease of white equipment, tele- and video products. The contract is normally for 12 months, though longer or shorter agreements are possible, then the products are returned to the company. Free installation, servicing, right to change the product for a new one.
Electrolux Ericsson	E2-home: intelligent living. Offers a number of relevant applications on a communication platform for homes designed for Intelligent Living. Service areas: Home Management, Family Management, Landlord Comunications
Cisco, Fastweb, Biticino, Ariston Digital, Pirelli Real Estate, Studio & Partners	Internet home. Domestic appliances can be monitored and managed from the Internet.
Deutsche Telekom	Virtual answering machine (T-Net-Box)
Toshiba 'home appliance rental packages'	Rental of an automatic washing machine, a two-door fridge-freezer, a flat television, and a cooking oven-range to those who live alone, such as students and workers away from home for about 30 $/month incl. any repairs, delivery, installation and removal

Table 2. Examples of new business models / product-service systems

as a partner provided a listing of washing expenses every month on the regular bill, using a smart electricity meter installed in the washing machine.

The indicative results of this field study are that families wash more conservatively, using the maximum load and changing the habits of laundering, resulting in positive effects on the environment by using less water, electricity, and detergents. However, the households seem to be hesitant towards functional sales instead of owning the appliances.

Co-Evolution of Technological and Belief Regimes

Coming back to the problems of transition management, it turns out that there are still substantial efforts to render in order to overcome innovation barriers resulting from behavioural lock-ins on the customer side. The transition theory provides some hypothesis on these lock-ins, highlighting the necessity of a co-evolution of technological, cultural, and behavioural belief regimes.[39] Obviously, the "pay per use" experiment did not convince customers to take advantage of this service on a broader scale. Reasons may be that the costs of conventional washing are not transparent to the customer, or that the field study did not address the proper customers (like larger households). These rather superficial explanations can be supplemented by a more detailed explanation reviewing the innovation barriers in terms of the "layers of transition theory". In this context, *Shove* [23] demonstrates that the transition of laundry habits depends on a particular relationship and interplay between several regimes determining why, how, and when laundry is done and explaining, how novelties unfold between as well as in societies in the sense that they form a new trajectory:

- belief regime:
  - manufacturers' definition of cleanliness evolve over time, like "absence of bacteria", "whiteness", inducing a transition in habits. For example today less than 7% of the washing in the UK is done at 90°C [23] but also requiring other detergents operating at lower temperature levels (= 'system-in-system' innovation);
  - social norms arise by developing informal rules over time (like Frank Sinatra's Song "I did it my way" suggests), i.e. habits evolve over time. For example, regardless of contamination certain laundry is washed in a certain frequency (beds, pyjamas, cushion covers, etc.);
- technological regime:
  - availability of both, washer and dryer, set the scene for inducing new rules and practices ('automatical' selection of certain clothes to wash and to try by habit regardless of contamination).

The "loss of ownership", insofar, refers to a social norm established as an informal rule in modern societies over decades. Thus, the proprietary thinking,

[39] See [23], p. 3.

practised as a guideline of individual career aspirations over years, turns out to be the most resistive lock-in in the implementation of the system innovation "pay per wash". Unless the belief regime and thus people's attitude towards necessary conditions of everyday life is not changed with respect to ownership, functional sales of a kind "pay per wash" will not disseminate. Since the creation of a new set of 'common rules' in the belief regime needs a long time to create (in the sense that occupancy is more important than ownership), the deployment of new business models, tested in micro level technological regimes like the Gotland experiment, will not be accepted short-term in society.

The experiences with the development of new business models indicate that there is still a long way to go, since corresponding business development strategies are not routinised yet, i.e. general decision moments cannot be defined, the circle of departments to include varies from model to model, the costs of setting up new business models are still difficult to calculate, etc., and – despite common expectations – sustainability effects of system innovations like PSS cannot be satisfactorily evaluated due to problems in methodology and data availability. What seems to be clear is that the implemention of new business models, like PSS, beyond the B2B area, has to be followed by a comprehensive systemic development process, managing a co-evolution of technological, institutional, organisational, behavioural, and cultural regimes.

6 Managing Transition in the EEIS: Problems and Suggestions

This paper runs the risk of being far more optimistic in the evaluation of sustainability progress in the EIIS than other opinions. The production of electrical and electronic devices is one of the fastest-growing industries. The global players in the EIIS are now starting to explore new business models in an attempt to open up markets for the 4.6 billion people (75% of population) in the developing countries who consume only 25% of the earth's resources [40]. Especially new business models (e.g. leasing, rental), which rely on small initial investments only, are promising in developing countries. Nevertheless, the increased use of ICT in developing countries will have an overall negative environmental impact. Most of the 58% of the energy increase projected by the US Energy Department by 2025 is expected to be used in the developing countries.[41] On this basis, it must be assumed that the waste management problem, despite of implementing less resource intensive EEE, will increase dramatically over the next 20 years.

Against this background, the transition process will only lead to a really sustainable electronics industry, if the shifts in technological, belief, and gover-

[40] See `http://www.cfsd.org.uk`, version of May 9th 2004.

[41] See `http://www.uspolicy.be/issues/energy/energy_usgreports.asp`, and `http://www.uspolicy.be/Article.asp?ID=050C02C5-9CC1-4517-ACB4-325F0B5CD394`, version of May 12th 2004.

nance regimes are carefully inter-related. This is – analysed in this paper as one of the overriding problems – supposed to be the main task for an improvement of the transition process in the EIIS.

First of all, there are indications of transition barriers, resulting from a malfunctioning co-evolution of the regimes up to now:

- The present regulation framework seems to partly hinder the development of new business models. When considering additional legislative obligations in a specific innovation system to increase the steering effect of an initial policy instrument, policy makers should be aware of existing technological or cognitive trajectories, path dependencies, and the current industry development. Cumulative intensity and timing of regulation have to be evaluated, since two or more regulations lined up in the same innovation system may be mutually indifferent, complementary, additive, or even conflictive. Within the governance regime of the EIIS the evaluation of the cumulative intensity of the regulation context remains unclear. So does the question whether the portfolio facilitates or hinders the development of sustainable business models. Insofar, the governance regime is only fairly prepared to modulate beliefs and technological choice towards sustainability at this point.
- In other areas of the governance regime, e.g. educational policy or R&D policy, the timing of measures to promote the development of more sustainable business models does not fit to the ongoing shift in technological regimes. For instance, research up to now is far too much focussed on technological aspects and too little on integrating socio-ecological and economical aspects. Unfolding technological diversity does not sufficiently go hand in hand with sociological investigations of shift in the society's technologies.
- The technological regime shift is not sufficiently coordinated with the changes in belief regimes. As stressed in Chapter 5.3, the shifts in consumer needs obviously follow shorter innovation cycles than the shifts in ownership habits, in design ethics or principles, and in fundamental business strategies ('time-to-market' versus 'multi-cascade innovation systems with longer life cycles'). Obviously, the governance approaches of promoting education about sustainability are somewhat late.
- The international nature of the transition is neither sufficiently tackled in the research arena, nor in the transition management. Since it may be assumed that the EU will become more and more important in environmental and sustainability policy making, the problems of transnational or even global innovation systems (like the EIIS) and their transition within the multi-layer governance structure of EU – nation states, federal structures within nation states (regional and local level)– remain unresolved in transition management and transition theory. In addition, the co-evolution of different belief regimes, reflecting international diversity in cultures, and otherwise distinct socio-technical regimes and landscapes is not treated

sufficiently up to now. Especially, the diffusion of EEE products in different societies under the normative direction of sustainability may need a decisive co-ordination of different belief regimes. This is referred to as the relation between horizontal and vertical structuring of system innovation. Of special interest is also the role of standardisation in the co-evolution of technological and belief regimes.[42]

- Within the technological regime evaluation methods to assess the sustainability effects of new business models are missing. In that sense, the extension of indicator systems is desirable to also record behavioural changes towards sustainability. Also, the impacts of increasing transaction costs in setting up new business models have to be evaluated more in depth.

To sum up: a vast transition barrier is constituted by the only moderate co-evolution of the transition regimes. This can even be broken down to the level of *niches* as the place of putting innovations, like new product-service systems, into practice. A comprehensive business model comprising new use patterns, ownership models, and the involvement of new innovation actors, needs to be co-ordinated like a bicycle chain – you may turn it only all at once. In view of these transition barriers, selected recommendations may be drawn for the practical transition management of the EIIS and for further research in transition theory:

(1) The transition of the EIIS towards sustainability now needs a particular manner of lock-in management [26], which has to care for keeping in line with the unfolded technology regime and their embedded dialogue, strategy, and tools.[43] The pathway to resources protection, energy saving, prolongation of use and life cycles, recovery of resources, and intelligent ways of need satisfaction must not be deserted. The EuP directive may be a lever to initiate a lock-in management between the EU and member states, fostering and consolidating the implementation process of eco-design requirements for energy-using products, because it will further promote life-cycle thinking with all innovation actors. More trans-national networks, compound of industry and research agents, have to be set up to learn about further implementation of the regulatory framework (like a RoHS network, an EuP network) since the mutual implementation benefits of joint action in industry are expected to be very high.

(2) Research on transition management and transition theory has to cover different aspects related to the:
    - timing of transition management: when to start, how to set up a transition arena, whom to involve, how to impose system change and how to detect time windows for a paradigm change?
    - instrumentation of transition management: what to do in particular phases of the transition (choice of instruments like technology assess-

[42] See [23] on these topics.
[43] See again Table 1.

ment, delphis, scenarios, funding programs, standardisation, cluster-management, etc.)?
- target regimes of transition: how to encourage bottom-up activities like networks on a local and regional level, involving research on social and institutional aspects of transition[44]; and how do institutions and behaviours change?
- co-evolution aspects of transition: how to set up links for an interaction of technology, belief and governance regimes, what are the indispensable systems of interaction?
- management processes: how to enhance coordinate governance action between the parties involved (question of how to coordinate people, who do not coordinate their activities at all)?

A study of historical transition processes using the concepts of the transition theory would give more insights into the dynamics and steering capability of transition. Especially, the dynamics of regime changes and the role of technology (enabling or disabling role), the role of belief and value systems, and the importance of governance should be investigated in ex-post analyses.

References

1. INVERSI (2004): Internalization versus Internationalization, A Framework of Action for National and International Environmental Policy against the Background of Increasing Globalization and the Development of Electronic Markets, Final Report, August 2004, edited by RWI – Rheinisch-Westfälisches Institut fuer Wirtschaftsforschung, Essen.
2. Kemp, R./Loorbach, D. (2003): Governance for Sustainability Through Transition Management, paper for EAEPE 2003 Conference, November 7-10, 2003, Maastricht, The Netherlands, (http://meritbbs.unimaas.nl/rkemp/Kemp_and_Loorbach.pdf), version of May 1st 2004.
3. Konrad/Scheer (2004): Systeminnovationen: Begriff, Fallbeispiele, Nachhaltigkeitspotenziale; http://www.ioew.de/dienstleistung/publikationen/Vortrag%20Wilfried%20Konrad%20und%20Dirk%20Scheer.pdf, version of April 30, 2004.
4. IÖW (2001): Politische Strategien für eine nachhaltige Dynamik sozialökologischer Transformationen, Studie für das BMBF, Berlin.
5. Rotmans, J./Kemp, R./van Asselt, M. (2001): More Evolution than Revolution – Transition Management in Public Policy Foresight. Vol. 03, no. 01, feb. 01, p. 005, (http://www.icis.unimaas.nl/publ/downs/01_12.pdf), version of May 2nd, 2004.
6. Berkhout, F./Smith, A./Stirling, A. (2003): Socio-technological Regimes and Transition Contexts, ESRC Sustainable Technologies Programme Working Paper, Number 2003/3, June 2003, p. 3 http://www.sustainabletechnologies.ac.uk/PDF/Working%20papers/FB1.pdf), version of May 1st 2004.

[44] See especially [5], p. 15.

7. Jacob, K. (2004): Governance for Industrial Transformation – The Scope of the Challenge. in: Jacob K., Binder M., and Wieczorek A. (eds.) (2004): Industrial Transormation between Ecological Modernisation and Structural Change, Environmental Policy research Centre, Berlin
8. Linscheidt, B. (2000): Kooperative Steuerung als neues Modell der Umweltpolitik – Eine theoretische Einordnung, in: Staatshandeln im Umweltschutz – Perspektiven einer institutionellen Umweltökonomik, Hrsg. Kilian Bizer, Bodo Linscheidt und Achim Truger, Neue Folge Band 69, Berlin.
9. Wegner, G.: Entstaatlichung der Umweltpolitik durch innere Institutionen?, in: Formelle und informelle Institutionen – Genese, Interaktion und Wandel, ed. by G. Wegner and J. Wieland, Marburg 1998, p. 35 – 68
10. Jacob, K. (2004): Politikexperimente mit ungewissem Ausgang, in: Ökologisches Wirtschaften. Hrsg. vom Institut und Vereinigung für ökologische Wirtschaftsforschung, No. 2 2004, p.14 (translation by the author)
11. Elzen, B. (2003): Transition to Sustainability through System Innovation – Summary Report from Workshop and Follow-up Activities, (http://sustsci.harvard.edu/events/twente02_transition_ws+followup.pdf), version of May 2nd 2004.
12. Wieczorek, A./Vellinga, P. (2003): The Need for Industrial Transformation. in: Jacob K., Binder M., and Wieczorek A. (eds.) (2004): Industrial Transormation between Ecological Modernisation and Structural Change, Environmental Policy research Centre, Berlin
13. Meyer-Stamer, J. (2003): Understanding the Determinants of Vibrant Business-Development: The Systemic Competitiveness Perspective, (http://www.mesopartner.com/publications/Systemic_WIRAM.pdf), version of May 2nd 2004.
14. Rotmans, J. (2003): Transitions and Transition Management; http://www.oecd.org/dataoecd/56/15/2487244.pdf version of April 30th, 2004
15. Wegner G./Pelikan, P. (2003): Evoluationary Analysis of Econommic Policy. Cheltenham, Edward Elgar.
16. Ashford, N. (2003): Conzeptualizing Pathways for Sustainable Transformations, Conference Paper presented at the Berlin Conference on the Human Dimensions of Global Environmental Change 5-6 Dezember 2003, (http://www.fu-berlin.de/ffu/akumwelt/bc2003/download.htm), version of May 1st 2004.
17. Rennings, K. et al. (2003): Blueprints for an Integration of Science, Technology and Environmental Policy (BLUEPRINT), (http://www.blueprint-network.net/pdf/atticonvegni/blueprint.pdf), version of May 1st 2004.
18. Steurer, R. (1999): Politik und Psychologie für den Umweltschutz. Politologische Betrachtungen zu einem transdisziplinären Thema, (http://www.eco.psy.ruhr-uni-bochum.de/ipu/literatur/rundbrief/nr9/diskurs_steurer.html), version of May 1st 2004.
19. Hafkesbrink, J./et al. (1998): Abschätzung der innovativen Wirkungen umweltpolitischer Instrumente in den Stoffströmen Elektroaltgeräte/Elektronikschrott, Untersuchungen des Rheinisch-Westfälischen Instituts für Wirtschaftsforschung, Heft 26, Essen.

20. Stirling, A. (2004): Diverse Designs: Fostering technological diversity in innovation for sustainability, Paper submitted to BMBF/ESRC/RIW/EC conference on 'Innovation, Sustainability and Policy', Kloster Seeon, Germany, May 2004.
21. Hafkesbrink, J. /et.al. (2003): ECOLIFE II – Eco-efficient Life Cycle Technologies – State-of-the-Art Technology Report in the Electronics Industry Innovation System.
22. Nightingale, P. (1997): The Organisation of Knowledge in CoPS Innovation, Paper prepared for the 7th International Forum on Technology Management, Kyoto, Japan, CoPS Publication No. 14.
23. Shove, E. (2002): Sustainability, System Innovation and the Laundry, Published by the Department of Sociology, Lancaster University, Lancaster LA1 4YN, UK, at http://www.comp.lancs.ac.uk/sociology/papers/shove-sustainability-system-innovation.pdf, version of May 7th 2004.
24. Dalhammer, C. (2004): Integrated Product Policy and Product Chain Innovation: The Role of Legislation and its Interaction with other Policy Instruments, Paper submitted for the international conference on Innovation, Sustainability and Policy, 24-25- May 2004
25. SUSPRONET status report, AREA 2 (2004): Product-Service Systems to information users, p. 26 at: http://www.suspronet.org/fs_reports.htm, version of May 9th 2004.
26. Faber, A./Rood, T./Ros, J. (2003): Evaluation of Early Erocesses in System Innovation, at: http://www.fu-berlin.de/ffu/akumwelt/bc2003/download/Faber_full.pdf, version of May 12th 2004

An Example of a "Managed Transition": The Transformation of the Waste Management Subsystem in the Netherlands (1960-2000)[1]

René Kemp[2]

[1] The contribution is based on the MERIT-ICIS project "Institutional change in the transition of waste management in the Netherlands" for the NWO research programme "Milieu en Economie". It draws on contributions of Derk Loorbach and Saeed Parto.

[2] United Nations University, Maastricht Economic and social Research and training centre on Innovation and Technology (UNU-MERIT), Keizer Karelplein 19, NL-6211 TC Maastricht.
r.kemp@merit.unimaas.nl

The contribution by Joachim Hafkesbrink deals with a very interesting issue: transition management in the electronics industry innovation system. It is about an issue in which I take a great interest as someone who developed the concept of Transition Management together with Jan Rotmans for the Dutch national government.[3] The issue concerns *the extent to which system innovations and transitions offering sustainability benefits can actually be managed by public decision-makers through public policy.*

In his contribution, Hafkesbrink says that the transition towards recycling of electric and electronic equipment waste was managed through two acts of European legislation, the WEEE and RoHS Directives. Rather than discussing transition management in the electronics industry innovation system about which I know little, I would like to offer a discussion of a 'managed transition' about which I do know something, namely the transformation of the Dutch waste management subsystem in the 1960-2000 period. It is an interesting story of a transformation process which was managed through various acts and through a newly created organisation, the Waste Management Council (Afval Overleg Orgaan, AOO), which acted as a change agent and mediator but, as I will argue, could only do so because of special circumstances described in the contribution (acute waste management problems at a time of waste scandals when there was agreement about the waste management hierarchy). In this

[3] The approach of transition management is described in [4, 6, 7, 8, 12, 13].

contribution I take issue with the idea that a transition can be managed by a policy act.

During the past 40 years, the Netherlands experienced a transformation in waste management: from uncontrolled landfilling (waste dumping) towards a differentiated waste-handling system of recycling, incineration with energy reuse, and controlled landfilling. It is unclear whether this transformation has ended — changes at the European level (the disappearance of waste borders) may lead to further change (even backwards) — which is why I will talk about transformation and not about transition.

In some ways the transformation meant a return to the old practice of recycling. 150 years ago, recycling was a common practice in the Netherlands: glass, metals, old fabrics, and certain types of organic waste were being collected by individual traders ([10]). At the end of the 19th century, such activities became less economical, and more and more private entrepreneurs stopped collecting waste. The "schillenboer" with his horse collecting shells of vegetables no longer exists. Waste collection became a public task handled by municipalities.

Most of the waste (including rising quantities of chemical waste) was being landfilled; a small part was reused or incinerated in newly built incinerators. In 1912, the first incineration plant was opened in Rotterdam, while Amsterdam and Leiden followed in 1918 and 1914 respectively. In Den Haag, in 1918 a small incineration plant was opened which even generated electricity on a small scale. The incinerators were built in urbanized areas lacking landfill sites in the vicinity.

Waste was also used for filling swamps and ditches ("slotenrijden") to generate new land for settlements. No track was kept of the types of waste having been disposed. The Netherlands basically had an uncontrolled waste management subsystem, in which waste was disposed of with few environmental considerations. The principal issue was to get rid of waste.

In the 1970ies, waste and unsustainable waste management practices received increasing attention: concerns were raised about how waste was being managed, problems arose with creating new landfill sites (because of local resistance), and the 1972 report to the Club of Rome followed by the oil crisis in 1973 put attention to scarcity of materials.[4] Waste disposal was increasingly seen as a problem.

Special legislation for waste was passed and responsibilities were given to provinces. With the introduction of the *Hazardous Waste Act* (1976) and the *Waste Act* (1977), the Dutch provinces received the planning and coordinating tasks, while the implementation, to a large degree, remained with (cooperat-

[4] For Ad Lansink, the inventor of the waste management hierarchy (which became widely known as the "ladder of Lansink"), a direct link between raw materials and energy existed. As he said: "The Club of Rome report really established this link for me because it talked about a shortage of not only raw materials but also of energy. I felt that waste was potential raw material for energy generation." (interview with Lansink by Parto, Feb. 17, 2004).

ing) local authorities (collection and disposal). The reason for this change in responsibilities was the intention to put an end to the (uncontrolled) dumping on landfills and to benefit from economies of scale for incineration.

Provincial borders were closed for waste transports and the operators were given the exclusive right and obligation to collect waste in a certain region. Operators were guaranteed necessary supply (*processing certainty*), and transporters had a guaranteed demand. The activities were organized as municipal service, controlled by local politicians officially in control, responsible for funding (from `www.aoo.nl`). Central to policy thinking was the "waste hierarchy" proposed in the parliamentary motion of Ad Lansink in 1979. The waste management hierarchy covered the path from prevention, through re-use (of products), recycling (of materials), and incineration (with energy-production) to landfilling as the last option. The motion became law in 1986 and was an important cognitive institution ([11]). From the late 1970ies on, waste was increasingly seen as "a waste of resources" in polity. Business also started investigating ways to reduce waste as part of its environmental policy.

To reduce the volumes of waste for disposal, the Dutch government opted for a *differentiated waste-stream* approach in which certain types of waste (notably paper and glass) were singled out for recycling. The initial reluctance to adopt the separate waste system came from the municipal waste-collecting services that had to change their practices. Other actors, like NGOs and private businesses, performed new activities such as the collection of paper and glass. The systematic collection of the bulk of recyclable waste and organic materials would only become institutionalized in the 1990ies ([11, p. 7]).

Despite these attempts for upgrading waste practices, many activities in the area of waste management still suffered from their small-scale nature and from inadequate environmental protection. For example, up until the 1990ies, soil protection measures were absent in virtually all landfills and flue gas scrubbing in waste incineration facilities was inadequate (from `www.aoo.nl`). There was considerable political and community resistance to the construction of new landfills and incineration plants, with the resistance reaching a high peak in the 1980ies, following the discovery of leaking landfills (Vogelmeerpolder) and contaminated land (Lekkerkerk and Griftpark). Waste scandals were a frequent news item in the 1980ies. The two most important ones were: Lekkerkerk, in which it was discovered in 1980 that new houses had been built on soil containing chemical waste which had been landfilled, and Lickebaert, where in 1989 dioxins (coming from incinerators of AVR and AKZO) were discovered in the milk of grazing cows. Five waste incinerators were closed because of dioxin emissions and at least one plan for a new landfill (Does in Leiden) was abandoned because of opposition. Whilst capacity was decreasing, waste volumes kept growing, leading to capacity problems. In 1991, as a result of lack of regular waste management capacity, it even became necessary to store waste in push barges.

At the end of the 1980ies, the Dutch waste management system was in a state of crisis. The system was reviewed by the Landelijke Coördinatie Commissie Afvalbeleid (Commissie Welschen) in 1989 which concluded that "the current organisation is fragmented, dispersed, and small scale". It argued for the creation of a nationally oriented organisation for disposal, to manage overall waste volumes and keep disposal costs under control. For incineration, but also for organising waste management from cradle to grave (chain management), four waste regions (encompass several provinces) were envisaged, each with three to four million inhabitants (from `www.aoo.nl`). This advice led to the appointment of the AOO through the Co-operative Agreement for Waste Disposal VROM/IPO/VNG (1990). The AOO would play an important role in the modernisation of the waste system.

From the beginning there were problems with the four waste-regions system. Municipalities wanted to sign contracts with waste companies in other regions and, because of capacity problems, waste had to go to other regions for incineration. In 1996, upon the advice of the Commission Epema, it was decided to centralise the responsibility for waste control at the national level. The legal basis for the centralisation is the last amendment of the Environmental Management Act that came into force in May 2002. Especially efficiency considerations fostered this decision. The centralization was welcomened by new private collecting and transport companies which wanted to operate nationally.

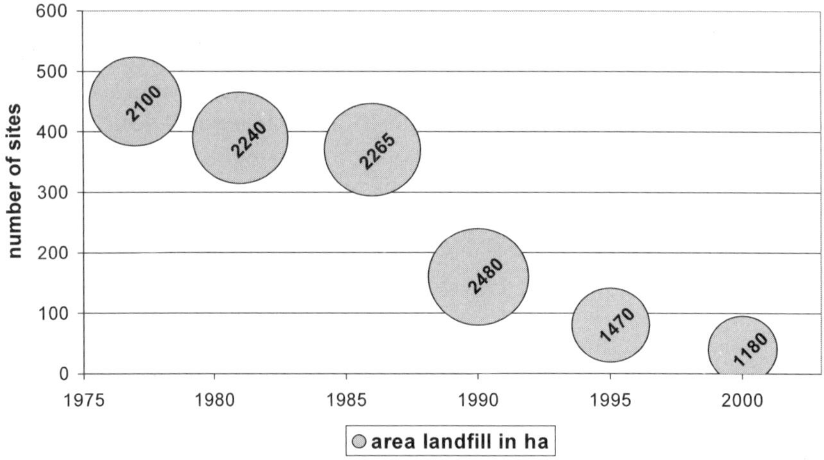

Fig. 1. Source: AOO

In the 1977-2000 period, the number of landfill sites fell from 450 to 40 (an eleven-fold decrease) thanks to the differentiated waste-stream approach and targeted policies (such as the packaging covenants), the landfilling ban of 32 waste streams, and steadily increasing costs for landfilling, creating an incentive to move up on the "waste ladder". The amount of waste being landfilled fell from 14 Mtons in 1990 to five Mtons in 2002, a reduction of nine Mton. Today, all landfills have advanced systems of soil protection and methane extraction. Meanwhile, the capacity of incineration increased gradually, from 2.2 Mtons in 1980 to 4.9 Mtons in 2000. Between 1995 and 2000, incineration capacity increased by 2 Mtons. Recycling almost doubled between 1985 and 2000 from 23.5 Mtons to 45.3 Mtons of material.

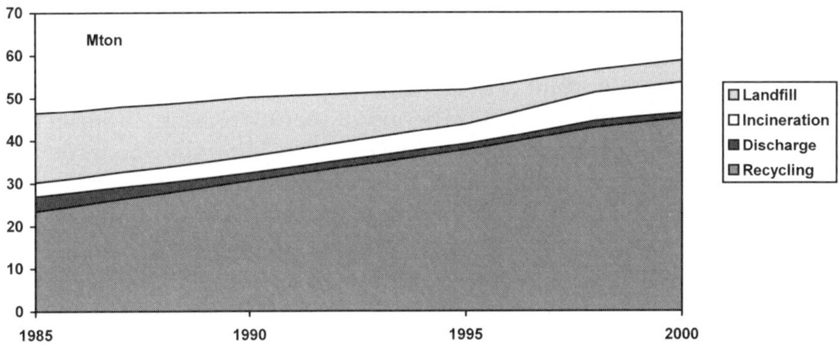

Fig. 2. Source: AOO

Multilevel Interaction Processes

The transition to a system of recycling and increased incineration with controlled landfilling as a last resort option is often viewed as the result of policy. Such a view, although not wrong, overlooks that policy itself was the result of various changes: the growing volume of waste, the waste scandals in the 1980ies and early 1990ies, and changes in beliefs (such as the belief that waste is "a waste of resources" and that landfilling should be done in a hygienic manner and only be used as a last-resort option) in a period in which environment was very much on the people's mind. The waste scandals helped to close down old incinerators and build better ones. Various waste acts provided the basis for policy and the AOO, created in 1990, brought together the three layers of government (local, provincial, and central) to work in a joint policy network without clear legal status under an independent chairman.

The AOO played an important role in the transformation process. Negotiations between different layers of government and with private waste companies took place within the AOO, with the actors agreeing on the general direction of creating a modern and efficient system of waste management with less waste being landfilled. The environmental movement, while being officially opposed to incineration, did not create too much trouble because its supporters understood that high costs of advanced incineration systems necessitated a high tax for landfilling of burnable waste[5], which encouraged waste prevention and recycling. The waste companies welcomed the greater scale at which they could operate. For the AOO, the reorganization of the sector, with big companies from North America such as Waste Management Inc and BFI taking over small companies, was seen as a blessing. The big companies were committed to full compliance and had a strong incentive to respect the law.

A simple causality analysis disclosed that not a single driver was responsible, but that several drivers influenced each other. Packaging policies and the rising costs of waste management were influenced by other factors (growing waste volumes and opposition to landfills). Waste scandals (due to past waste practices) were an important aspect, allowing policy makers to modernize the waste management subsystem. Furthermore, the investments in incineration capacity are an important influence by necessitating a regular supply and policies to secure this (such as bans and a high landfill tax for burnable waste). Within the waste regime, the rule system and the roles of the different actors changed. Policy was thus endogenous, a response to immediate issues. To deal with problems of capacity, a new network organization (AOO) was created, which served as an important coordinator. The AOO is viewed by outsiders (such as Geelhoed in a speech at the AOO lustrum conference in 2001) as an example of the "poldermodel" of consensus-based politics, but the organization rather considers itself to be a change agent and mediator (Daemen and Huisman from AOO in an interview with Loorbach and Kemp, 7-9-04). The transformation can be viewed as successfully managed, but it also may be criticized for being overly expensive by relying so much on incineration and recycling ([5]).

Implications for Transition Management

What do we learn in terms of transition management? To me, this example teaches us three things. First of all, it shows that a transition or transformation cannot be controlled in any simple way. Different developments have to come together and to sustain each other. Secondly, it is useful to have a more or less commonly shared long-term orientation serving as the basis for coordination. Without this, policy can only react to immediate problems (act in a 'fire-brigade' fashion of putting out fires). Thirdly, since policy is problem-driven, you need acute problems for creating new institutions and for initiating changes which are helpful also for the longer term.

The idea of exploiting existing problems in a strategic way is a central element of the model of transition management outlined by Rotmans, Kemp, and Loorbach in various publications, which is currently being used in the Netherlands for managing transitions to sustainable energy, sustainable mobility, sustainable agriculture, and sustainable use of resources. The basic

[5] In 2002, the landfilling tax for burnable waste amounted to 79 euro per ton (62% of the price to be paid).

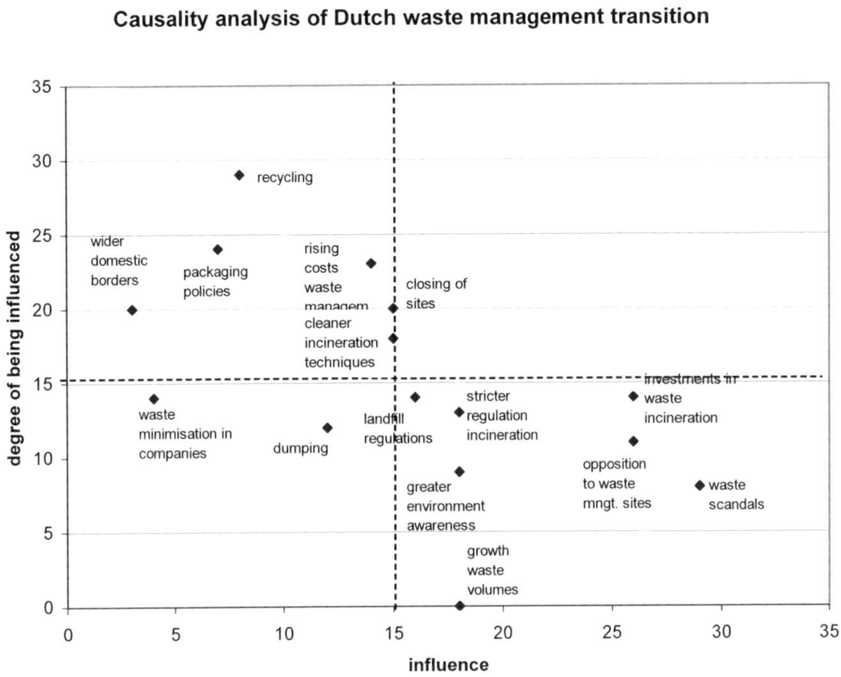

Fig. 3

steering philosophy is that of *modulation*, not dictatorship or planning-and-control. Transition management joins in with ongoing dynamics and is built on bottom-up initiatives. Windows of opportunity are exploited in a strategic manner. Transition management for sustainability tries to orient societal dynamics to participatorily defined sustainability goals for functional systems. Learning, maintaining variety, and institutional change are important policy aims.

The Dutch transition approach is innovation-oriented and very much bottom-up with long-term visions guiding societal experiments. To avoid lock-in adherence to certain paths, various paths are explored simultaneously. This makes sense given the uncertainty about the best option. In doing so, Dutch authorities rely on the wisdom of variation and selection processes rather than on the 'intelligence' of planning. Transition management is iterative and adaptive. A mechanism of self-correction based on policy learning and social learning is part of it. Through the various elements (programmes for system innovation, creation of transition agendas, the use of transition arenas) transition management offers a framework for policy integration, helping different ministries to collaborate. Whereas other countries are engaged in managing transitions in an implicit way, the Netherlands do so explicitly. The commitment to transition allows for cooperation between ministries, but also for political choices which are needed for leading production and consumption closer to sustainability. It is not a substitute, but a new framework for politics.

References

1. www.aoo.nl
2. Afval Overleg Orgaan (2002): Het poldermodel op de afvalhoop?: De rol van overleg in het toekomstige afvalbeleid? AOO 2002-03
3. Daemen J. (2003): Waste Management Planning in the Netherlands, ppt presentation at waste management council meeting on 5-6 November 2003
4. Dirven J., Rotmans J. and Verkaik A.-P. (2002): Samenleving in Transitie. Een vernieuwend gezichtspunt, LNV, ICIS en Innovatienetwerk Groene Ruimte en Agrocluster, April 2002
5. Dijkgraaff E. (2004): Regulering van de Nederlandse afvalmarkt, proefschrift EUR, Rotterdam
6. Kemp R. and Rotmans J. (2001): The Management of the Co-Evolution of Technical, Environmental and Social Systems. paper for international conference Towards Environmental Innovation Systems, 27-29 September 2001, Garmisch Partenkirchen, Germany (forthcoming in Weber M. and Hemmelskamp J. (eds.) Towards Environmental Innovation Systems, Springer Verlag)
7. Kemp R. and Rotmans L. (2002): Managing the Transition to Sustainable Mobility. paper for international workshop "Transitions to Sustainability through System Innovations", University of Twente, 4-6 July 2002, (forthcoming in Boelie Elzen, Frank Geels and Ken Green (eds.): System Innovation and the Transition to Sustainability: Theory, Evidence and Policy, Cheltenham, Edgar Elgar)
8. Loorbach D. and Rotmans J. (2004): Managing transitions for Sustainable Development. In: Wieczorek A.J. and Olsthoorn X. (eds): Industrial Transformation – Disciplinary approaches towards transformation research. Kluwer, The Netherlands, Forthcoming
9. Loorbach D., Parto S., and Kemp R. (2003): From Waste Disposal to Waste Management: Transitions in Waste Management in the Netherlands. mimeo, Maastricht
10. Loorbach D. (2003): A short history of waste in the Netherlands, mimeo
11. Parto S., Loorbach D., and Kemp R. (2003): Institutional Change During Transitions: The Case of the Dutch Waste Management Sector. Paper presented at the IHDP Meeting October 16-18 2003, Montréal, Canada
12. Rotmans J., Kemp R., Asselt M.v., Geels F., Verbong G. and Molendijk K. (2000): Transities & Transitiemanagement. De casus van een emissiearme energievoorziening. Final report of study "Transitions and Transition management" for the 4th National Environmental Policy Plan (NMP-4) of the Netherlands, October 2000, ICIS & MERIT, Maastricht
13. Rotmans J., Kemp R. and Asselt M.v. (2001): 'More Evolution than Revolution. Transition Management in Public Policy', Foresight 3(1), 15–31

Comment: Management of Industrial Transformation: Potentials and Limits from a Political Science Perspective

Klaus Jacob

Environmental Policy Research Centre, Freie Universität Berlin, Ihnestraße 22, D-14195 Berlin.
jacob@zedat.fu-berlin.de

In spite of considerable efforts since the foundation of modern environmental policy 30 years ago, industrial production is far from being environmentally sustainable in many sectors and regarding many issues. Despite the decoupling of environmental degradation from economic growth for a few types of pollutants, many environmental problems remain unresolved and even more new problems become apparent. Climate change, the loss of biodiversity, the overuse of water and land, and the release of harmful chemicals are only some examples of this trend. The consumption of natural resources and the utilisation of the environment as sink for emissions exceed an acceptable gauge for the long-term. Hence, an encompassing structural change in industry is necessary. Market mechanisms on their own are not sufficient to bring the necessary change about. The temporal and geographical horizon of both consumers and producers of industrial goods is too short-sighted to translate external effects which originate from production and consumption of goods into adequate prices. Still, existing political endeavours have proved to be insufficient to cope with such a task as well. The management of an encompassing structural change in industry is a task which is disputed within the political system itself. Such an objective potentially conflicts, at least in a short-term perspective, with the objectives of safeguarding employment and economic growth.

The fierce battles between the different economic actors, environmental agencies, and also within the government itself, concerning the introduction of ambitious policy measures for environmental protection, indicate such a political difficulty. The examples by Joachim Hafkesbrink on transition management in the electronics industry innovation system and by René Kemp on transitions in waste processing in the Netherlands show, that far-reaching and long-lasting transition processes actually do take place. However, there are no examples of a coherent, comprehensive, and enduring management of such processes. Up to now, transition processes which led to system innovations were not subject of a strategic steering process.

Against this background, the concept of transition management has been developed and tested, in order to overcome current deficits in governing industrial transformation (see for the following: [8, 3, 5, 4]). The concept focuses on "system innovations" which are understood as a fundamental change of technical, social, regulative and cultural regimes which, in their interaction, fulfil specific societal needs such as those for transport, food, housing, water, or energy. A system change requires the co-evolution of technologies, infrastructure, regulations, symbols, knowledge, industrial structure, etc. Historical examples of system innovations are the transition from wind-powered to steam-powered ships, from wood-based to coal based-energy production, etc. Such changes typically require a timeframe of 30-40 years.

Such a long timeframe and the necessary encompassing changes are not manageable by conventional governmental steering. Traditional policy-making is sectionalised in specialised departments and, like business actors, rather short-sighted.

It is therefore proposed to institutionalise a transition management which should provide advanced performance in steering system innovations. However, this does not include any claim to actually plan transitions, but instead to influence the direction and the speed of transformation processes. It is suggested to divide the process in four distinct phases: 1) the creation of an innovation network (transition arena) for a defined transition problem which includes representatives from government, science, business, and NGOs. Initially, such a network should not be larger than 10 to 20 persons. 2) The generation of integrated visions and images about possible transition paths which span 25-50 years. Based on these visions, intermediate objectives should be derived. 3) The execution of experiments and concerted action according to the transition agenda. Experiments may thereby refer to technologies, regulations, modes of financing, etc. 4) The monitoring and the evaluation of the process and the respective implementation of the emerging learning processes. Successful experiments should be taken up by the normal policy process and their diffusion should be promoted.

Within such a framework, a transition policy should be developed which 1) stimulates variety in order to avoid lock-in situations, 2) is integrated and legitimised by the conventional decision-making mechanisms, 3) takes place with a long-term perspective even if no immediate successes can be achieved, and 4) is coordinated with the different levels of policy formulation. Transition policies can fall back on the existing instruments of R&D, environmental, and sectoral policies, but their application has to be integrated and coordinated.

Up to now, there is no example for success or failure of such efforts for a transition management. However, since 2001 projects are under way in the Netherlands to experiment with such a type of strategy.

What expectations can be associated with these proposals? Are strategies for transition management suitable to give transition processes a momentum towards sustainability? The need for a long-term policy is not doubted facing the environmental problems of current production and consumption patterns. However, it can be questioned, whether at all and how a rather small "transition arena" will be able to formulate such objectives which receive attention and will be respected in political practice.

To have any practical outcome, the visions developed in such a framework have to represent a consensus among the stakeholders. However, in this case it is not very likely that such objectives do provide any governing effects since they will be very much inclined to the status quo and will not harm the interests of any major stakeholder.

In case more substantial objectives are developed, restrictions can be expected similar to the impediments scientific advisors face in the policy process. Discourse and persuasion undoubtedly do have their own effects in political decision-making. However, the mode of bargaining among different interests is largely ignored by the concept of transition management, while the importance of visions is largely overrated. The binding character of visions is rather low and, on top of this, there is no reason to presume the existence of a single vision. Instead, there are a number of competing visions about a desirable future. The use of nuclear power is just one example for this: while some policy makers expected a bright future connected to the introduction of nuclear power with cheap and almost limitless energy availability, others were afraid of accidents or acts of sabotage against nuclear power plant's protection. Imagining a transition management for the introduction of nuclear power, the actors did not agree on a simple unique vision of a desirable future. The same holds true for the phase-out of nuclear power.

The limitations of persuasive policy-making has been sketched inter alia by Sabatier and Jenkins-Smith [9]. In their model of policy change, they distinguish between deeply rooted core beliefs, policy beliefs, and secondary beliefs. Changes cannot be expected regarding the core beliefs of actors, and hence, to achieve policy change against fundamental values and interests, arguing and convincing are not likely to lead to success.

Moreover, the selection of actors who should take part in a transition arena is not substantiated. Should actors get included who have an interest in keeping the status quo? Which motives to take part in the development of a transition strategy can be expected from the different actors? Blockades can be anticipated in case the bargaining should not decrease the ambitiousness of the objectives. Cooperation in the transition arena is not the only possibility for the involved (and the affected) actors to accomplish their interests. The usual channels for pursuing and mediating their interests in the political system remain available and they cannot be expected to promote common welfare. Up to now, the proposals for transition management do not envisage the identification of such a common welfare.

The call for a more comprehensive policy integration, which covers industrial, innovation, and technology policies, is likewise not novel. Since the emergence of modern environmental policy, a cross sectoral integration of environmental concerns is called for, up to now, however, with only moderate success [2]. The governmental departments comply with different objectives and follow different rationalities in their action. The demand for far-reaching and enduring policy integration has been flatlined facing the institutional rigidities of the political process.

Furthermore, the call for keeping a diversity of options during the experimentation has to be assessed critically. On the one hand, it is necessary to avoid suboptimal lock-in situations. Furthermore, governmental actors do not dispose of the necessary information to decide about suitable technologies. Still, on the other hand, keeping options open causes costs which can be considerably high if the technologies at stake are capital-intensive and close to market introduction, or do require a particular infrastructure. At a certain point in time, it is necessary to decide about the investment in one or another technology.

The emphasis on niches as a starting point for system innovations neglects other sources of transformation. The opportunities for an endogenous change are thus underestimated [11]. The authors suggest a typology distinguishing the source of resources for the renewal of systems and the degree of coordination regarding the allocation of these resources. The purposive transition, as suggested by the Dutch proposals for a transition management, a transformation based on external resources and a high degree of coordination, is just one out of four possibilities. Other forms of transition may be the reorientation of trajectories (with the system's resources and a low degree of coordination), an endogenous renewal (with the system's resources and a high degree of coordination), or emergent transformation (with external resources and low coordination). Each type of transition requires a different steering approach, respectively is open to different modes of intervention.

Despite the provisos against the opportunities for an encompassing management of transitions, a focus on system innovations remains helpful. The analysis of the interplay of different actors in the value chain and on policy patterns rather than single policy instruments opens up a perspective on new points for intervention and allows the analysis of impediments and restrictions of policy innovations [1]. The broad view on systems allows an identification of the determinants of the ecological performance in product chains and an identification of the different actors involved, their interests, and their resources enabling them to comply with new demands or to oppose new policies. The weakest part in the chain may be identified in such an analysis (see also [12].

Furthermore, the broadening of the temporal perspective allows new opportunities for a strategic alignment of policy measures. Again, in this respect, new points for intervention might be identified. The emergence and diffusion of innovation is a non-continuous process, and phases of radical technological change alternate with phases of incremental change. During phases of intensified innovation development, new technologies can more easily be established on the market, compared to relatively stable phases. The political impulses which are required for a successful diffusion of environmental innovation can then be weaker than during phases of stability [7]. Such windows of opportunity can be prepared by the promotion of alternative technologies. However, the proposals for time strategies do not provide answers to questions such as which technologies to promote and how to identify a window of opportunity ex ante. This entails practical problems: Is it advisable for the current environmental and technology policies to wait for decentralised fuel cells for combined heat and electricity production? Or should current policies promote the already available motor-driven mini CHP's, or should it bet on another technology, e.g. microturbines? Keeping all these options open simultaneously is likely to be too costly. Furthermore, the prospects of future technologies is, against the background of the urgent current problems of environmental degradation, no excuse to wait with the exploitation of the potentials of currently available environmental technologies.

References

1. Jacob K. (2004): Governance for Industrial Transformation. The Scope of the Challenge. in: Jacob K., Binder M., and Wieczorek A. (eds.) (2004): Industrial Transormation between Ecological Modernisation and Structural Change, Environmental Policy research Centre, Berlin, 7–20
2. Jacob K. and Volkery A. (2004): Institutions and Instruments for Government Self-Regulation: Environmental Policy Integration in a Cross-Country Perspective. Journal of Comparative Policy Analysis: Research and Practice 6 (3), 291–309
3. Kemp R. and Rotmans J. (2005): The Management of the Co-Evolution of Technical, Environmental and Social Systems. in: Weber/Hemmelskamp (2005, 33-55)
4. Kemp R. and Loorbach D. (2003): Governance for Sustainability Through Transition Management. EAEPE 2003 Conference, 7–10 Nov. 2003, Maastricht
5. Loorbach D. (2002): Transition Management. Governance for Sustainability. Paper presented at the Conference Governance and Sustainability. New challenges for the state, business and civil society, 30 Sept.–1 Oct. 2002, Berlin
6. Loorbach, D. and Rotmans, J. (2006): Managing the Transition to Sustainable Development. In: Olshoorn X. and Wiezorek A.J., Understanding Industrial Transformation: Views from Different Disciplines. Springer, Heidelberg, 187–206
7. Nill J. (2004): Time Strategies of Transitions and the Transformed Role of Subsidies as Environmental Innovation Policy Instrument. in: Jacob K., Binder M., and Wieczorek A. (eds.) (2004): Industrial Transormation between Ecological Modernisation and Structural Change, Environmental Policy research Centre, Berlin, 255–307
8. Rotmans J., Kemp R., and Asselt M.v. (2001): More Evolution than Revolution. Transition Management in Public Policy. Foresight 3 (1), 15–31

9. Sabatier P.A. and Jenkins-Smith H.C. (1999): The Advocacy-Coalition Framework: An Assessment, In: Sabatier P.A.: Theories of the Policy Process. Boulder, Colorado
10. Sartorius, Ch. and Zundel (eds.) (2005). Time Strategies, Innovation and Environmental Policy. Edward Elgar: Cheltenham, UK; Northampton, MA.
11. Smith A., Stirling A., and Berkhout F. (2005): The Governance of Sustainable Socio-Technical Transitions. In: Research Policy, Volume 34, Issue 10, 1491–1510
12. Spaargaren G., Mol A.P.J., and Buttel F.H. (2006): Governing Environmental Flows. Global Challenges to Social Theory. MIT Press, Cambridge

Part II

Innovations and Sustainability

Leading Innovations to Sustainable Future Markets

Klaus Fichter[1] and Reinhard Pfriem[2]

[1] Borderstep Institute for Innovation and Sustainability, P.O. Box 37 02 28, D-14132 Berlin.
fichter@borderstep.de

[2] Carl von Ossietzky University of Oldenburg, Faculty II, Chair for Strategic and Environmental Management, D-26111 Oldenburg.
reinhard.pfriem@uni-oldenburg.de

The economic problems which presently affect more or less all industrialized countries point beyond the conformity crises which we have seen in the past 150 years of factory society. These conformity crises were most of all characterized by technological key innovations which became the fundamental innovations leading to a new level of prosperity: steam engines, steel, chemicals and electronics, petrochemicals, and automobiles [13]. According to these Kondratieff cycles, named by Schumpeter after the man who first came up with the idea, we are, with the fifth long wave (in the form of the information society), in the midst of changing circumstances of economic organization. New information and communication technologies, the globalization of the economy, most notably in terms of organizational and social aspects, the culturally charged consumer demands in a growing number of fields, recursive of company offerings – all of these have implied consequences for companies and business competition which are new when compared to the previous 150 years of factory society.

The United Nations Conference on Environment and Development in Rio de Janeiro in 1992 was not only focused on overcoming global ecological challenges, but also aimed at promoting the economic and social development. This was done to create business, labor, and lifestyle models (especially in industrialized countries) which are just and transferable on the entire planet and between generations.

Social stability as part of economic structural change is, in this respect, an important element of sustainable development. In this sense, the activity of intermediary service companies such as the Hamburg firm Projektwerk e.g. during the SUMMER Project[3] is not sustainable simply because paper is saved by using electronic means of communication, but for an entirely different reason: In view of the socio-economic and organizational aspects of current structural severances, it is of tremendous importance to bring and to network together knowledge-intensive service companies in such a way which promotes new entrepreneurship and new structural elements of sustainable economies.

However, future company success is not about different structures, but about new search processes: The globalized companies of the 21st century are faced with the challenge of achieving (in addition to the established classical process and product competition realms) the generation of new markets as a third level of economic competition [11, 9]. How can sustainable future markets be actively developed by means of companies and company networks, and which strategy processes, methods, and instruments can effectively be of support? With an eye on the central question of the SUMMER project, the following will present selected insights.

Co-Evolution: Sustainability Through Innovation and Cultural Change

The empirical findings from SUMMER show how many companies still limit themselves to the (technical) dimension of product and process innovation. As a rule, the dominant focus on technical optimization and problem solving in the field of innovation dismisses rebound and growth effects as the "side effects" or "long-term consequences" of technically efficient solutions. Thus, in the debate on sustainable development, the deeper-seated reasons for non-sustainability are (rightfully so) referred to, which are linked to consumption patterns and lifestyles and therefore to behavior and values. In the context of innovation, sustainability consequently represents first and foremost a cultural challenge and demands a debate on the communicative reach of sustainability innovations or *alleged* sustainability innovations. A sustainable development therefore requires a tight interaction of company innovation and a change in culture and behavior. From this the notion is derived that success-oriented innovation management and marketing must be aware of these factors. A culturally aware and interpretative management [10] purposefully reflects not

[3] The research project "SUstainable Markets eMERge" (SUMMER) was funded by the German Federal Ministry for Education and Research and was carried out from April 1, 2001, to May 31, 2004. Research partners were: Carl von Ossietzky University Oldenburg (project leader Reinhard Pfriem), the Borderstep Institute for Innovation and Sustainability, Berlin, the Institute for Product Durability Research, Ecco, Ltd., as well as two large companies (BSH Bosch and Siemens Appliances, Ltd. and MohnMedia), two young, small companies (Velotaxi, Ltd., Berlin, and Projektwerk, Ltd., Hamburg), and two company networks (Joiner Network KonnexX and the Institute for Building and Living, Ostfriesland). The results are availaible at `www.summer-net.de`.

only its own values and beliefs but also those of customers and partners as well. An ability to connect culturally and to set ideas, technologies, or business concepts in motion is needed here. Creativity is understood to be an interactive, social process whose progression requires continuous identification and (re)interpretion of values and assessments. This concretely means:

- Integration of lifestyle research when conceptualizing solutions for consumers and "looking beyond the horizon" of the selling point, taking the usage of products into account,
- forecasting changes in customer behavior and values which might result from the planned product, service, or system innovations, as well as
- the examination of cultural acceptance and the ability to activate new solutions by means of incorporating lead users [8, 12] and of testing them as part of interactive user tests and systematic market research.

The Generation of Sustainable Future Markets as Innovation Process

"Sustainable future markets" are new markets (or markets which will appear in the future) which, through their traded products or rendered services, contribute to the aims and targets of sustainable development, namely based on a definable system unit (product life cycle, usage system, demand field). The emergence and make-up of sustainable future markets represent an economic reorganization process. Here, companies can contribute by means of product, process, and service innovations as well as through brand-new, or expanded forms of trade and intermediary functions. For companies, this signifies the generation of new markets as a third level of economic competition [11, 9] beyond known and established classical realms of process and product competition. Under which prerequisites are product, process, service, or system innovations bound to the emergence of new markets or market segments? As the empirical studies from the SUMMER project show, three types of new markets and market segments can be distinguished:

New Markets Through New Demands and Function Groupings

Product or service innovations can establish a new demand or a new desired function. For example, the demand for photovoltaic facilities and components is made possible through solar technology innovations and by means of innovative business models. New government regulations (e.g. take-back and reutilization of recyclables) can be the basis for a new demand (take-back, remanufacturing and recycling of materials) and can establish a new market. Additionally, new markets emerge through the recombination of existing products and services to form new functions. An example for this is the company Velotaxi Ltd., Berlin (`www.velotaxi.com`, participants in the SUMMER project) which took the rather separate services of bicycle rickshaws and mobile advertising spaces and recombined them together with a modern and appealing vehicle design, thus creating a new system service.

New Markets Through Extended Transaction Realms

Some service or product innovations basically establish new forms of trade transactions or a fundamental expansion of the trade realm. Here, it essentially means innovations in the field of information and communication technologies and service innovations which use internet technologies for novel, forward-leading platforms of electronic trade. Examples for this are the internet service company abebooks.com, which took the previously local trade in used and antique books to the national/international level, or Amazon and eBay which both created a forum for extensive trade in used items. The new aspect is not the traded goods, but the trade realms and the sell-buyer interaction.

New Markets Through Innovative Fulfillment of Existing Demand

Through product, service, and system innovations the way to meet existing demands can be fundamentally changed. The function to be fulfilled remains unchanged. However, it is now achieved through a changed product manufacturing or product usage in a new way. Examples for this include leasing, sharing, pay-per-use, and reuse models, all of which lead to a changed use of physical products.

Search and Discovery Pathways for Sustainability Innovations

A study of 68 examples from the business world as part of the SUMMER project [3] shows six typical search and discovery pathways for sustainability innovations:

1. *Sustainability as the dominant goal to be achieved by the innovation process:* The starting points of this emergence path are demand and sustainability problems (e.g. the overfishing of fish stocks) which key players, as visionary entrepreneurs, governments or non-governmental organizations perceive as unacceptable. Meeting demands or removing grievances as an explicit contribution to a sustainable development form the main goal to be aimed at and help shape a sustainable solution over the entire course of its realization.
2. *Sustainability as an integral corporate goal and strategic success factor:* As opposed to the first path of emergence, sustainability does not represent a dominant, highest-priority objective, but is rather integrated as an important and formally equal element into a corporate-political goal. Here,

the corporate-political aspect precedes the innovation process. Sustainability is seen by the relevant power promoters as a strategic success factor, represents a normative specification, and is checked and reflected upon by various methods and instruments (e.g. life-cycle assessments) over the course of the innovation process.

3. *Sustainability potential as a "coincidental" discovery in the ongoing development process:* Whereas with the first two emergence paths explicit sustainability goals accompany the innovation process from the start, sustainability considerations in this case bear fruit only after the development process has begun. Participants "discover" and/or realize over the course of a development process that the desired solution would make a recognizable contribution towards sustainable development.
4. *Sustainability standards as a possible correction of the ongoing innovation process:* As in the case just described, sustainability aspects appear in the consciousness of the innovating actors only after the innovation process has begun. Still, as opposed to the "coincidental" discovery of a positive potential, sustainability considerations here play a prominent and success-relevant role due to public criticism, stakeholder pressure, and a lack of enforceability.
5. *A retroactive attribution of sustainability and its use as a sales argument:* Yet another way towards sustainability innovation is found in those innovation processes during which sustainability requirements or goals have not played a noteworthy role. In these cases, it is realized in hindsight, i.e. upon market introduction or even in the course of diffusion, that the product or service also has environmental advantages.
6. *Sustainability as an "invisible hand":* In the sixth and last path of emergence, sustainability aspects are found in the consciousness of the innovating actors neither before, during, nor after the innovation process. A sustainability contribution seems to be created through the "invisible hand" of the given public policy and technological conditions. A potential for sustainability or a realized contribution towards sustainable development is perceived only by external observers (scientists, etc.). For example, electronic market places which facilitate trade of used consumer or investment goods can be interpreted as a contribution towards extended product usage.

Process Competencies for the Generation of Sustainable Future Markets

Meeting sustainability challenges in the innovation process brings new challenges to the visionary abilities of companies and to corporate foresight. It also makes it necessary to handle the complexity of life-cycle and system observations, to determine social and ecological side effects, to manage complex actor networks and stakeholder demands, and to deal with conflicts which might emerge between economic, ecological, and social objectives. Which components do companies have to develop in order to be able to integrate and strategically weave sustainability requirements into the innovation process? The SUMMER results show that two company competencies are central here: first, the ability to define the "realm" of the innovation management in such a way that, in spite of competition and time pressure, sustainability requirements and chances can be brought to bear and the "desire and ability" can be supported in a way conducive to sustainability (context management). Second, the capacity to systematically shape the communication processes and personal encounters needed for a reflexive strategy development and successful creation of actor cooperation.

Corporate Context Management

The concept of context management assumes that the emphasis of planning and control must be shifted (with increasing dynamics and complexity) towards the creation of enabling requirements and suitable innovation contexts. A goal-oriented influence upon the terms of the innovation management not only means the establishment of appropriate search and selection rules, but the making of the resource accoutrements of innovation projects (among other things) dependent on their sustainability contribution as well. Concrete starting points of context management are:

- Establishment of a corporate mission statement for sustainable development (vision, company values, basic principles of sustainability, code of conduct, etc.)
- Corporate governance standards for the innovation process (sustainability as an integral R&D goal; environmental, health, and/or safety requirements as obligatory verification criteria when evaluating ideas; research results; product and service concepts; etc.)
- Establishment of sustainability-oriented dominant logics (e.g. sustainability-relevant influence factors as "given" in trend monitoring, scenario management, and technology roadmapping; orientation towards the overall life cycle of materials in the evaluation of ideas and/or solutions; expansion of horizons to usage systems; etc.)

Actor Interactions: Dynamic Formation of Cooperation and Networking

One of the most striking features of the SUMMER project and the 68 examined innovation examples was the phenomenon of actor cooperation. The finding that actor cooperation plays a central role in innovation processes and in the development of sustainable problem solutions is not new. What is new, however, is a dynamic view of the formation and changes of cooperation and network processes. While the research on cooperation had structural questions like trust, complementarity, and power relations in the foreground, a dynamic view places aspects of actor interactions at its focal point. Regular strategy meetings between network partners, innovation workshops, stakeholder dialogue meetings, or the close and early bonding of pioneer customers and lead users are forms of interaction which have (not coincidentally) gained in meaning. Actor interactions fulfil central functions of knowledge and interest integration against the background of rising markets, technology dynamics, and the increasing division of labour in the innovation process. For example, the following forms of actor integration had a central significance during the international market development for bicycle taxi services in the Velotaxi company (studied as part of the SUMMER project):

- Regular (moderated) strategy meetings
- Presentations and discussions with authorities, politicians, etc.
- Meetings of network partners moderated by an external "coach"
- Innovation workshops with important cooperation partners and lead users
- Active networking of cooperation partners and the creation of a Velotaxi community
- Counseling and support of founders and new customers by Velotaxi Ltd.
- Test runs, exchange of experience, interactive user research
- Team development workshops

Interpreneurship: Interactive Methods in Company Dealings

The experiences of the SUMMER project equally emphasize, like the empirical case analyses, that the integration of sustainability requirements in the innovation process and the active generation of sustainable future markets through companies and company networks bring a necessary attitude change in the debate on management concepts and instruments. While the research and discussions on innovation and sustainability management to date have been dominated by information and analysis instruments and organizational directional considerations, the SUMMER results emphasize the necessity of placing interactive concepts and instruments at the center of attention.

The entrepreneurial role (entrepreneurship) during the initiation and implementation of new solutions can be seen from a new interactive perspective in the light of the central meaning of actor interactions. This interactive change in viewpoint leads to a new concept of entrepreneurship, which we call "interpreneurship". With reference to Schumpeters idea of "creative response" [15] we define *interpreneurship as the entrepreneurial creation of new (inter-)connections for the discovery and realization of innovative solutions* [5, p. 326]. The creation of new inter-connections is essential on two levels: first, in the supplier structure, which, compared to before, no longer can be

Innovation Process	Instruments and Methods		
	Analysis Collecting Information and Facts, Analysis, Evaluation	**Organization** Structuring and Regulating Processes and Systems	**Interaction** Creating Meetings, Dialogue, and Cooperation
Orientation Recognizing problems and needs Sustainability-oriented direction of strategies and search fields	• Trend Monitoring • SWOT Analysis	• Sustainability Visions • R&D Objectives	• Stakeholder Dialogue • Strategy Coaching in Companies and Company Networks
Generation Sustainability-oriented inspiration when generating ideas, Initiative for sustainability-oriented innovation projects	• Publication and Patent Analysis • Technology Portfolios	• Incentive Systems for Employees • Eco Design Parameters	• Business Innovations Workshop • Sustainability Roadmapping in Multi-Actors-Networks
Acceptance Reflexive selection, Determination of side effects and integration of "sustainable lead users"	• Capital expenditure evaluation • Eco-Efficiency Analysis	• Staged Gate Process (e.g. Dow) • Lists of acceptable materials and substances	• Scenario Workshops with Companies and Stakeholders • Lead User Integration (e.g. Customer Workshops)
Realization Creation of directional certainty during market preparation, Production configuration and market introduction	• Customer Surveys • Environmental Performance Indicators	• Codes of Good Practices • Quality Criteria for Product Labelling	• Interactive User and Test Market Research • Marketing Cooperations

Table 1. Methods and instruments in innovation processes (selected examples)

understood in terms of a single company, and second, called for by the interactive change in viewpoint, the economic interaction between vendors and customers. As we saw with the SUMMER project, customers can definitely initiate sustainability innovations. Furthermore, customers also influence the development of new products and services as well as their market introduction. Customer involvement is a fundamental component of innovative sustainability policy.

For the practical realization of the interactive change in viewpoint (as we call it), the SUMMER project developed and tested a wide array of various methods and instruments (see Table 2). The use of these methods contributed greatly to the success of the various projects.

Method/Instrument	**Application Example**	**Effect**
Business Innovations Workshop	Velotaxi GmbH Berlin, Germany	Initiation and Planning of a New Generation of Vehicles (Zero Emission Passenger and Delivery Vehicles) (Market Introduction 2006)
Customer Workshop	MohnMedia, Gütersloh, Germany	Initiation of a Development Project for Paper-Saving Print Formats (Substantial Reduction of Paper Refuse)
Strategy Coaching in Company Networks	Joiner Network KonnexX, Baden-Württemberg, Germany	Development of a Sustainable Furniture Program (Market Introduction 2004), New Marketing and Network Management Strategies
Pioneer Customer Integration	MohnMedia, Gütersloh, Germany	Realization of an FSC[4]-Certified Print Product (Mail Catalogue); Broadening of Market for Print Products from Sustainable Forestry
Interactive User and Test Market Research	BSH Bosch and Siemens Household Appliances GmbH, Munich, Germany and `www.leshop.ch`, Switzerland	Development of Demand-Suitable Service along the Refrigerated/Frozen Goods Chain, Intelligent Inventory Management and Online Grocery Ordering
Network Coaching	IBW Company Network for Ecological Construction and Living, Ostfriesland, Germany	Development of an Extensive Offering of Seminar and Teaching Services on the Topic of Healthy Living and Ecological Building as Part of a Market Development Strategy of the Company Network

Table 2. Interactive methods of the SUMMER projects (selected examples)

Using the Lead User Approach for Sustainability Innovations

In the search for the triggering forces and impulses in the innovation process, [8] developed the "lead user" concept. Hippel determined, as part of his empirical studies on innovation processes, that the manufacturer in no way does alone profit from a product or process innovation. Deliverers or customers can also be the main beneficiaries of a new solution to a problem, and thus can become active as co-innovators in its development and implementation: "(1) Lead users face needs that will be general in a marketplace – but face them months or years before the bulk of that marketplace encounters them, and (2) lead users are positioned to benefit significantly by obtaining a solution to those needs." [8, p. 107]. As part of the SUMMER practical project, lead users could be integrated in different phases of the innovation process, e.g. through interviews or dialogues (orientation phase), innovation workshops (generation phase), or through active participation in user tests and interactive test market research (acceptance phase).

Hippel's lead-user approach is a very valuable basic concept, but it has to be adjusted to the specific requirements of sustainability-oriented innovation processes (see Figure 2 below). When starting a lead user project (step 1) one of the most important aspects is that the search field, which has to be selected in the beginning, has to have a pressumably high potential to contribute to sustainable development. Search activities should be guided by a question which reflects unmet demands or severe problems in the context of sustainability. For example, in the SUMMER project with BSH Bosch and Siemens Household Appliances, search activities focussed on the fact that each year the amount of food, which decays in refrigerators of German households, equals the value of 5 billion Euros. This is not only a waste of money, but also a waste of natural resources. Thus, the leading question in search activities and idea generation with BSH was: How can the handling of food, which needs to be cooled, be made more comfortable and controllable for consumers in order to reduce the spoilage of food and the waste of money and natural resources?

Step 2 of the lead user project aims at identifying relevant trends and needs regarding to the selected search field or target market. Whereas lead user projects carried through so far have mostly been limited to the analyses of market and technological trends, sustainability orientation makes it necessary to also have a close look at relevant trends in society, political developments (new laws, etc.) and business-related questions regarding the state of the natural environment (climate change, emission trading, shortage of drinking water etc.). This broader look follows the idea of multiframing [2]. Multiframing is based on reframing the views of managers and on generating new insights and ideas by combing different perspectives on the world and on future markets. To include sustainability aspects in trend analysis leads companies, for example, to the recognition of the fact that nearly two thirds of the world population are poor, in many cases denied access to proper services, water, health, and, above all, the opportunities to improve their economic and social outlook. For this reason, a growing number of member companies of the World Business Council for Sustainable Development follows a concept of doing business with

---

[4] Forest Stewardship Council (`www.fsc.org`).

the poor in ways which benefit the poor and the company at the same time [19].

After step 3 (identification of lead users and their ideas) and step 4 (developing solutions and concepts with lead users) it is essential that the generated ideas or concepts are evaluated not just with regard to realizability, market potential and profitability, but also referring to effects on sustainability (effects on health, safety, livelyhood, and the natural environment), cultural trends and behavioural aspects (lifestyle, etc.). The incorporation of lead users into pilot projects and test market research opens the opportunity to get to know usage needs and behaviours to identify cultural, behavior-relatedand adaptive abilities or the barriers to new problem solutions, and to achieve rapid results following the philosophy of learning by doing/learning by using. This is how restraint and implementation risks, as well as potential unwanted side effects and rebound effects, could be determined. The lead user concept is suited not just for the development of new solutions for usage and demand, but also (in its extended form) as a learning and control instrument regarding to the sustainability effects of innovations.

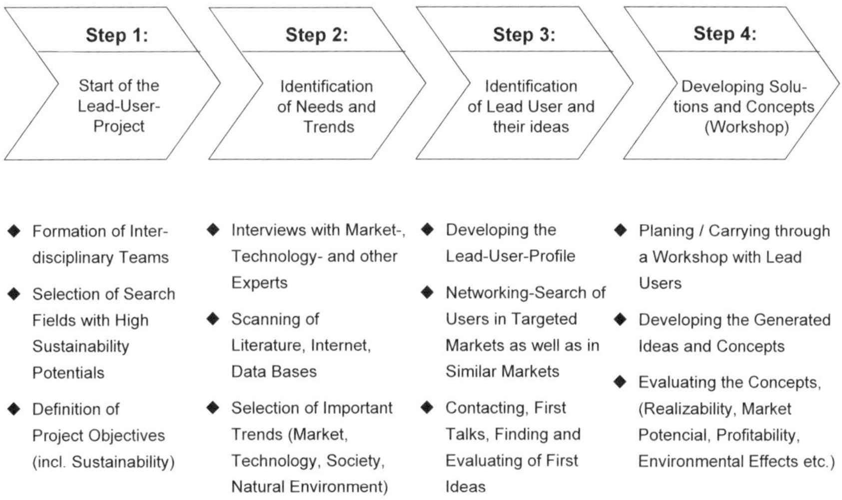

Adapted from Herstatt, Lüthje, Lettl 2003, p. 62.

Fig. 1. Implementation of a sustainability oriented lead-user project

From Lead User to Lead Market

In conclusion, the results of the SUMMER project, which concentrated on the emergence and implementation of new solutions up to the point of market

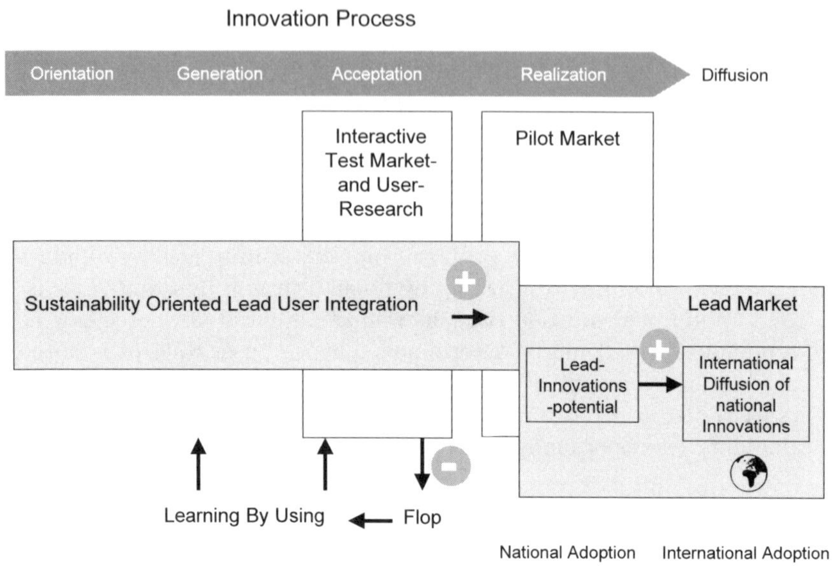

Fig. 2. From lead-user integration to lead-markets for sustainability

introduction, can be brought together through the concept of lead markets, which deals with the international diffusion of national/regional innovations [1]: "Lead markets are regional markets (normally countries), that use a certain innovation design and have specific features (lead market factors) at their disposal sooner than other countries. This increases the possibility that other countries will, on a wide scale, adopt the same innovation design". "Innovation design" is understood here as the technical specification of an innovation idea (e.g. the European standard frequency system for cellular phones). Following a selection process in a regional market, a design comes to the forefront which the authors of [17] call the "dominant design." However, the regional adoption of an innovation (pilot market) does not automatically mean that this innovation is implemented internationally. Its diffusion depends on internationalization factors, such as export orientation, transference/transferability of demand conditions, and government incentives.

The lead-user approach developed and tested in real-world situations as part of the SUMMER project, and its interactive user and test market research can systematically be connected to the concept of lead markets and the international diffusion of environmental-friendly innovations. By doing so it is possible to trace back lead markets to the very first innovation idea and its search and discovery conditions. Thus, a new research and formation field is defined which identifies considerable (world) market potentials while also being able to contribute to global and temporally transferable production, consumption, and lifestyle patterns.

References

1. Beise M. et al. (2003): The Emergence of Lead Markets for Environmental Innovations. In: Horbach J., Huber J. und Schulz T. (Hrsg.): Nachhaltigkeit und Innovation, Rahmenbedingungen für Umweltinnovationen. München, 11–53
2. Bolman L.G. and Deal T.E. (2003): Reframing Organizations, Artistry, Choice, and Leadership. 3rd ed., San Francisco
3. Fichter K. and Arnold M. (2004): Nachhaltigkeitsinnovation [Sustainability Innovations, Sustainability as a Strategic Factor]. Oldenburg, download at www.summer-net.de
4. Fichter K. and Paech, N. (2004): Nachhaltigkeitsorientiertes Innovationsmanagement [Sustainability-Oriented Innovation Management], Prozessgestaltung unter besonderer Berücksichtigung der Online-Nutzung. Oldenburg, download at www.summer-net.de
5. Fichter K. (2005): Interpreneurship. Nachhaltigkeitsinnovationen in interaktiven Perspektiven unternehmerischen Handelns [Sustainability Innovations in Interactive Perspectives of the Entrepreneurial Role]. Marburg, Metropolis-Publishing
6. Gleich A.v. (1997): Innovationsfähigkeit und Richtungssicherheit [Innovation Capabilities and Directional Certainty]. In: Gleich A.v., Leinkauf S. und Zundel S. (eds.): Surfen auf der Modernisierungswelle? Marburg, Metropolis-Publishing, 245–261
7. Herstatt C., Lüthje C. und Lettl C. (2003): Fortschrittliche Kunden zu Breakthrough-Innovationen stimulieren [Stimulating Progressive Customers for Breakthrough-Innovations]. In: Herstatt C. und Verworn B. (Hrsg.): Management der frühen Innovationsphasen. Wiesbaden, Gabler-Publishing, 57–72
8. Hippel E.v. (1988): The sources of innovation. New York, Oxford, Oxford University Press
9. Heuskel D. (2001): Competition unlimited. John Wiley & Sons Inc.
10. Lester R.K. and Piore M.J. (2004): Innovation: The Missing Dimension. Cambridge, MA, Harvard University Press
11. Moore J.F. (1996): The Death of Competition. Leadership and Strategy in the Age of Buiness Ecosystems. Harper Collins
12. Morrison P., Lillien G., Searls K., Sonnack M. and Hippel E.v. (2001): Performance assessment of the lead user idea generation process for new product design and development. Working Paper, WP 4151, Sloan School of Management, Massachusetts Institute of Technology, Cambridge, MA
13. Nefiodow L. A. (1997): Der sechste Kondratieff [The 6th Kondratieff]. St. Augustin
14. Paech N. and Pfriem R. (2002): Mit Nachhaltigkeitskonzepten zu neuen Ufern der Innovation [With Sustainability Concepts to New Frontiers of Innovation]. In: UmweltWirtschaftsForum, 10. Jg., Heft, September 2002, 12–17
15. Schumpeter J.A. (1947): The Creative Response in Economic History. In: Journal of Economic History, 7, 1947, Nov., 149–159

16. Schumpeter J.A. (1949 [1911]): Theory of Economic Development. Oxford University Press
17. Utterback J.M. and Abernathy W.J. (1975): A Dynamic Model of Process and Product Innovation. Omega 3 (6), 639–656
18. Weizsäcker C. and Weizsäcker E. U. (1984): Fehlerfreundlichkeit [Error friendliness]. In: Kornwachs K. (eds.): Offenheit – Zeitlichkeit – Komplexität, Frankfurt, New York, 167–201
19. WBSCD – World Business Council for Sustainable Development (2004): Doing Business with the Poor – A Field Guide. Geneva

Comment: Sustainable Future Markets and the Formation of Innovation Processes

Klaus Burmeister

Z_punkt GmbH, The Foresight Company, Zeche Zollverein, Bullmannaue 11, D-45327 Essen.
burmeister@z-punkt.de

Sustainability is a model which must establish itself in the competition within a distinguished and individualised society. Sustainability does not succeed simply because it is economically, ecologically, socially, and culturally sensible and necessary for a future-oriented development. It can, however, succeed when sustainability innovations are properly connected with processes of change within economies and society, as during the fifth long wave (digitalisation), or in the transition to the sixth wave (health). Sustainability must renounce the concept of environment-conscious business, a thinking pattern that has become far too cramped. Saying goodbye to this ecological context, and taking on a new, common life perspective does, however, not mean giving up basic principles, but rather includes the acknowledgement of actual relations and power balances.

The generation and formation of sustainable future markets is the essence of the SUMMER project (Fichter and Pfriem give a detailed report on of the SUMMER project in their contribution in this volume, cf. also `www.summer-net.de`). The connection between innovation and sustainability during the transition to a knowledge-based society is here seen as a central principle. Moreover, the conceptual approach suggests (rightfully so) the common life and cultural integration of sustainability innovations. Sustainability as a "controlling principle" is in this regard much more than the possibility and/or realisation of a technological innovation. Identifying a technical innovation in the field of regenerative energies as sustainable may be relatively simple. Still, proving this with e.g. networked product and service innovations, such as mobile phones or desktop PCs, which are anything but long-lasting or energy-saving, is considerably more difficult. On the other hand, mobile phones can make a relevant contribution as individual gateway to the user as part of a regional mobility management. It is also evident that computers and the internet are rather sustainably expedient for building regional networks of exchange, as is the case with car-sharing as well.

The contribution of the SUMMER project is an extension of the context of a pure sustainability research to the conceptual approach including all phases of innovation processes. Here, the network effects of complex innovation processes spanning technology and practice are (necessarily) allowed for. "Search-and-discovery" pathways are offered which classify and evaluate sustainability innovations in the entire context of their creation and usage realms. Refraining from the construction of worn-out criteria systems for sustainability innovations further underlines the conceptual approach of the SUMMER project. The suggested procedural approach seems more plausible and practical, as it is often determined only during the process, or even ex post whether new sustainable future markets have developed, or not. The example of Velotaxi is an excellent demonstration of how the recombination of existing products and services can be re-deployed into new functional roles, in other words, how sustainable thinking can be turned around to achieve a breakthrough. Possible fundamental questions, regarding the kind of contribution advertising boards on bicycles can make towards sustainability, are omitted. It could be argued that Velotaxi has ultimately created more traffic in the city by increasing the touristic attractivity of Berlin. Such ponderings are of little help in the mostly contradictory world of practice. They are, however, important as they show how difficult a clear definition of sustainable markets is. Evaluating product and service innovation as part of their respective contextualisation therefore seems to be the more reasonable and promising way of doing things.

The overall promising conceptual approach of the SUMMER project (knowingly) enters new territory and is, in a few areas, (necessarily) ambitious. From the viewpoint of companies, highly complex and difficult-to-overcome requirements come about if they hope to meet the "fundamental requirements for a sustainability-oriented innovation management." As a rule, companies are not masters of the innovation domain. They are mostly themselves driven by outside forces, have to recognise technology options early and to adapt in a timely enough fashion for their innovation purposes, and, with this, should simultaneously keep all implications for sustainable markets constantly in focus. A self-sufficient "process competence for the generation of sustainable future markets" can therefore not yet be attested to companies. In a competitive situation characterised by time restraints, expanded added-value chains in form of networks, and an internationalisation of markets, the room for negotiation is limited even for sensitised companies which are open to the overall concept of sustainable development. In addition, sustainability innovations can often only be realised in cooperation with various partners. The "visionary power of the company" is often put to test due to political and and legal conditions they cannot influence upon. An accompanying socio-cultural discourse on sustainability (e.g. on sustainable consumption patterns) promoted by relevant social actors, and the continous introduction of sustainability thinking into political arenas, are still missing as a prerequisite for sustainable innovation processes. The SUMMER project is well aware of this situation and correctly features two decisive influential factors for a "directed" generation of sustainable future markets, namely context controlling and interaction. The question remains whether and how companies can play this game, and if (and how) the structurally and systemically disadvantaged sustainability aspect can be considered on an equal footing with other competing success factors in dynamic innovation processes of a wide variety of actors. This is not a conceptual question, but rather a socio-structural one.

There is no simple answer to how this can happen. A few approaches do exist, however. From the point of view of *corporate foresight* [1], a continued research perspective is a possibility and an opportunity for strategic partnerships. The concept of corporate foresight along with the SUMMER project combine (among other things) common innovation perspectives, the anticipation of future markets, and formation rights. Moreover, the applied methods show a high level of congruence (from trend monitoring to the "lead-user" approach all the way to scenario development). The management of innovations is interpreted as a core task for companies. According to the concept of corporate foresight, companies must continually translate the questions and needs of society into concrete solutions. Knowing this, successful innovations are developed at the intersection of scientific findings, social problems and needs, political requirements, and a multitude of difficult-to-calculate risks. Innovations require directions and guidelines for the application of this concept, too. Important here are competitive jobs in new, future-oriented industrial and service sectors, as well as a technological progress which takes the ecological system limits of the Planet Earth into account. The question is, whether and how the application of corporate foresight can be combined with the concept of sustainable future markets.

Corporate foresight creates a conceptual framework for systematic and continual strategic future work in companies. Corporate foresight can be interpreted as a corporate answer to the multitude of social, economic, ecological, technological, and cultural changes since 1989 such as the fall of the Berlin Wall, European unification, the advance of the Internet and mobile telephones, the rise and fall of the new economy, globalisation, 9/11, and the "shift to Asia." Over a span of only 15 years, the surrounding environment for companies as well as society has in some cases completely changed. Companies have reacted by (among other things) opening themselves to qualitative questions ranging from stakeholder dialogue, to corporate governance, all the way to corporate foresight.

Corporate foresight is understood as an approach towards development and qualification of innovation strategies. At the same time, corporate foresight seeks to develop an expanded innovation concept. The goal is to understand innovation as a medium- and long-term-oriented strategy, and thus, to establish itself on five innovation parameters, namely anticipation, quality, context, timing, and networks, all of which can be understood as a requirement for successful innovations. The question would be, whether and how a common approach can be developed from the common intersection of the conceptual undertaking. The charm of such an approach would also lie in the fact that both, the company-related as well as the environment-oriented approach, could each profit from mutually access. This way, the (to some extent) sobering experiences of the lead-user approach at Hilti or 3M and its continued development in practice could offer helpful suggestions for the sustainable lead-user approach propagated by the SUMMER project. On the other hand, the project-oriented exchange of practical experience could help refine even further the honed methodical instrument for the established practice in companies. Companies quite often lack an adequate interpretation of their problem areas in complex innovation processes. By the selection and new formation of instruments, and by their implementation in an empirically accompanied way, the intersections and differences of these two concepts would become obvious. Implied is a high degree of common problem areas and goals. The concept of "sustainable future markets" also offers, in this respect, an excellent basis for the development and funding of practically applicable transfer concepts.

References

1. Burmeister, K., Neef, A., and Beyers B. (2004): Corporate Foresight – Companies Create the Future. Hamburg

Directional Certainty in Sustainability-Oriented Innovation Management

Niko Paech

Carl von Ossietzky University of Oldenburg, Faculty II, Chair for Strategic and Environmental Management, D-26111 Oldenburg.
Niko.paech@uni-oldenburg.de

1 Innovation as an Ambivalent Mode of Change

What makes innovation representing a particular type of change, behaviour, or problem-solving, so ambivalent? An innovation's economic, ecological, and social effects reveal themselves once its use has begun. Since unintended side effects are discovered simultaneously with the creation of facts, it is always too late to reverse these effects. "Ambivalence is the experience we encounter when, just as we achieve or realize our goals, we discover that it is actually not the goal we had intended, but rather something else, up to and including its hindrance" [23, p. 80]. Two traits in particular, inseparable from the term "innovation," make it a double-edged phenomenon:

1. Innovation refers to a non-constant, non-linear mode of change, and a break with all things available and known, at least in terms of the context of the innovation at hand. Commensurate with the core question "How do new things come into being?", the problem solving potential of innovations is based upon expanding the pool of available solutions – regardless of whether they are new products, technical operations, organizational structures, institutions, etc.
2. Innovations require entering the realm of the unknown. They lack an exact prediction and direction, at least in terms of what we understand from traditional economic optimization, and comprise to consciously take risks which are associated with chances. The revelation of currently unknown options is hence best summarized as "No risk, no innovation!"

The ambivalent, or "paradoxical" structure of innovation arises from the fact that "innovations are reliant on conditions that cannot be fulfilled at the time of the innovation, as something completely new is generated. These are conditions that will rather need to be discovered, manufactured and tested during the innovation itself" [17, p. 14]. In addition to embedding the innovation object into the dominant usage context (seen in "conventional" innovations management primarily as a challenge), and its eventual marketability (recontextualization), which is unknown *ex ante*, an innovation process aiming at sustainability also requires dealing with another uncertainty, i.e. ecological and social side effects (sustainability for the future). This means not only the direct social and ecological effects of a marketable innovation, but also the indirect effects that could possibly stimulate growth, which could overcompensate for a long-term gain in efficiency or ecological consistency. "We have to deal with the paradox that technical innovations can, by solving known problems and fulfilling required needs, also generate new needs and previously unknown problems" [19, p. 149].

But to make the argument, that we should refrain from innovations, would be just as wrong as the constant appeal for a risk-taking mentality as a price for competitiveness and material wealth. From the point of view of sustainable development, it is much more important to complement innovation processes with configuration options which can lead to a decrease in typical modernization risks.

1.1 The Thin Line between Good and Bad Intentions: The Rebound Effect

Rebound effects can appear when a measure seen in an isolated context is established as having a positive sustainability effect, but also creates further effects on other decision levels, or other parts of the system which are seen to negatively influence sustainability. Three kinds of "rebounds" can be distinguished:

Technical Rebound Effects: The introduction of a new product, or process, which appears based on favourable sustainability principles can be seen as counterproductive from the point of view of another sustainability principle. For example, the automotive industry implemented a method for building lighter cars which led to considerable energy savings. The savings in weight were mostly achieved by substituting metal by plastics whose production and disposal can pose new ecological problems (efficiency advantage vs. consistency disadvantage).

Growth Effects: Sustainability innovations in the form of new products, processes, or usage systems can generate counterproductive growth effects if they do not lead to sufficient substitution of previous (less sustainable) solutions. The introduction of the 1.5 litre automobile could lead to many households acquiring this vehicle in addition to their existing "fleet" as a third car. The expansion of wind energy, or photo-voltaic use could induce additional resource and energy flows if the energy market absorbs the additional amount of regenerated electricity instead of accordingly reducing energy from fossil fuels and nuclear power.

Psychological Rebound Effects: Technical sustainability innovations can cause undesired reactions on the level of system use, thus cementing the exact consuming culture originally intended to be changed. Consequences, such as e.g. the introduction of the regulated three-way catalytic converter, which lately prevented an overdue societal confrontation with motorized individual traffic due to its "integrated alibi module," could be induced by the forthcoming series production of the so-called "fuel cell cars." Exactly the environmentally conscious people, who had until now chosen to not own a car, could now, as a result of such technical-ecological reassurances, become car owners. Additionally, car owners who had previously used their car only when there was no other alternative, would now possibly use their car for short "runs" as well.

1.2 Risk Effects

The difference between dangers and risks is, according to [10, p. 30–31], that the latter always represents the results of your own dealings or failures. [16, p. 55] defines two kinds of risk, of which the first is based upon well-known reasons and interrelations. Although its probability of occurrence can only be determined within a certain margin of error, the possible consequences are recognizable and can be reduced through "experience-based precautions".[1] The second kind of risk is characterized by a high level of uncertainty, where even the possible effects themselves are difficult to predict.[2] Similarly, but oriented towards a stronger methodology, is the typology of the German Government's Scientific Advisory Board for Global Environmental Changes (WBGU). It defines six kinds of risk, as shown in Table 1 on the next page[3]:

[4, p. 38] defines three kinds of risk, namely those whose potential for damage are:

- qualitative-punctual ("one strike"), i.e. due to extreme combinations of natural phenomena and very powerful technologies,
- dependent upon an extremely unstable condition of the system being encroached upon, or
- quantitative-cumulative ("little by little"), i.e. through a quantitative increase of individual, relatively harmless "nibbles" that come into being.

Special attention is paid to a possible ecological technology conflict resulting from the playing off the quantitative-cumulative against the qualitative-punctual problematic, in terms of a "efficiency revolution through risk technologies" [4, p. 32]. How accurate this estimate is can be seen e.g. in the use of thermal waste, gene technology, synthetic chemistry and (the return of)

---

[1] Example: The risk of a core meltdown.

[2] Example: While the possibility of scratching a transgenic plant from a test field is hard to predict, the determinability of the possible consequences of such a genetic transfer is rendered nearly impossible.

[3] See [15, p. 25f.].

Risk Types	Example	Possibility of Occurrence	Damage / Effects
Cyclops	Disease, Droughts, Volcanic eruptions	Unknown	Determinable
Pythia	Release of transgenic plants	Unknown due to unidentified biochemical processes	Unknown
The Sword of Damocles	Major Technologies: Chemicals, Nuclear Power Plants, Massive Dams	Low	High
Pandora's Box	Pumping the biosphere full of toxins due to uncontrolled expansion (e.g. DDT)	Unknown due to spatial and ecological complexity	Unknown
Cassandra	Creeping decay of ecological systems (e.g. climate catastrophes); Great time difference between cause and effect	High	Unknown
Medusa	Over-exaggerated dangers of ionized and electromagnetic radiation from cell phones	Low	Not scientifically provable

Table 1. Risk typology of the German Government's Scientific Advisory Board for Global Environmental Changes (WBGU)

atomic energy. After all, the high efficiency potentials of such technologies are seen by their proponents as contributions to sustainable development. Von Gleich [5] sees quantitative-cumulative risk scenarios as less problematic. In such situations, dealing with the unknown by applying a trial-and-error principle is adequate, at least when an effectual margin of error allows "something to go wrong" once or twice [5, p. 289].

But it should not be forgotten that it is exactly the combination of growth and quantitative-cumulative risk effects which can become a serious problem in sustainable development. All new economic activities imply an ecological price. New innovations, therefore, only earn the title of "sustainable" when the attained environmental savings or burden relief effects outweigh the "investment" of resources, energy, or other ecological costs incurred by the innovation's introduction. This also means: When something new is brought into the world and falls short of its envisioned sustainability effect, it automatically becomes part of the problem, because at its bottom line, it, too, has induced new material flows. This is an *ex ante* highly uncertain balance, therefore, a latent growth risk arises. Even if the balance turns out to be a positive one, an even more difficult problem remains: As the comparably more advantageous innovation achieves a gain in sustainability, it must replace the old, less advantageous solution. Otherwise, the innovation principle instead becomes an addition principle, therefore creating additional energy and material flows. Selection mechanisms might be missing, which in turn actually crowds out previously existing, less sustainable solutions. Even when these circumstances seem to primarily concern investment goods, they can also be relevant to consumer goods, particularly when the innovation (from the consumers' point of view) strongly differs from previous solutions.

On the other hand, another problem arises where available goods and operations are constantly replaced by new solutions due to effective selection: Intact, still useful components in the material sphere lose their value and wind up being thrown out. How can the danger of premature disposal and shortening of the usage and product life cycle be avoided? Here, the acceleration of innovation activities could lead to the cultivation of a "throw away" syndrome. Both scenarios taken together comprise a selection dilemma. The extremely (vague) solution could be: The substitution which is introduced must occur at the exact point in time when the usage life span of the solution to be exchanged has effectively reached its end. But this kind of uncertain undertaking can, in view of its frequency, add up to a largely underestimated risk. The latter is characterized by an integration of no less than three uncertain incidents:

1. Is the new solution at all more advantageous than the old one?
2. Will a substitution occur?
3. Will the substitution occur at the "right" point in time, or will it lead to a counter-productive depreciation?

2 Starting Point for Directional Certainty (Overview)

The often posed requirements for innovation processes with the goal of adequate directional certainty can be broken down to the following realms (among others):

- Limitation of the "effective power" [4, p. 35]; avoidance of technologies whose risks can turn entire generations into test objects;
- "Error leeway" [24]: The innovation being developed should be able to be corrected in case damage or dangers occur after market introduction;
- Reversibility: The innovation should not promote any "lock-in" effects or "structural conservatism" [19, p. 153].

The requirements allowing any consideration at all of such criteria are established at the start of the innovation process. Directional options available during the implementation phase of an innovation tie in with the previously established selection of an innovation object. Whether completed facts arise and which degree of freedom for a change of course or hindrance remains, subsequent to any damage potential occurring, depends on how far the concretion of the innovation objects was anticipated. Adaptation and formation boundaries, which continue to exist after the start of the process, require certain structural characteristics from the innovation process. Here, the four starting points should be considered, which are shown in Fig. 1 and form the subject matter of the following sections.

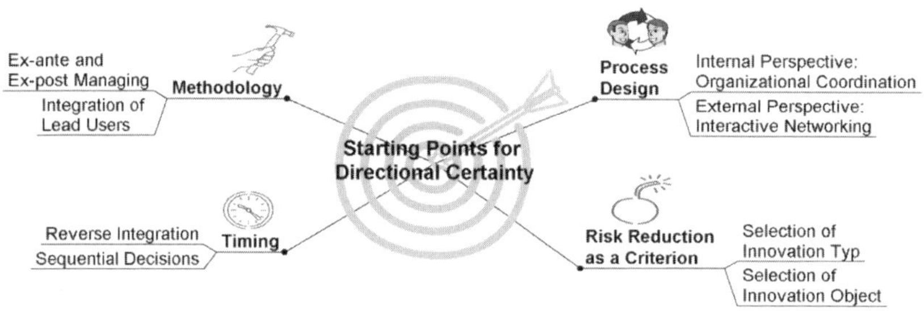

Fig. 1. Starting point for ensuring directional certainty

The process design of innovation projects can be, roughly simplified, represented by internal and external "guard rails." The first subsumes the influence of company-internal actors and measures of innovation management. Here, the organizational integration of innovation activity as well as the resulting innovation climate count. The second guard rail is based upon interaction with company-external actors who are integrated into the process. This differentiation should, however, not be misunderstood; the coordination of the external interactions is also, of course, a responsibility of innovation management. The following will draw special attention to risk reduction criteria as well as timing, due to its particular influence upon directional certainty. Parallel to this, the use of corresponding applied methodologies will be illustrated.

3 Risk Reduction Criteria

The selection of an innovation type spans the categories of product, procedures, service, usage system, organization, and institution. The actual innovation object is a moulding of the selected type. A product innovation in the automotive industry leaves e.g. the question open of whether the innovation object is meant with regard to an airplane, bicycle, or car and, – if a car is in fact meant – what it exactly means for the car. The type selection as well as its development into a specific object, which results at the end of the innovation process, both have great influence upon directional certainty.

In both parts of the decision, the avoidance of structural risks and rebound effects can enter in as additional selection criteria. Thus, an innovation idea, whose theoretically provable sustainability effect only represents a chance compared to its formidable sustainability risks, is best abandoned in favour of an alternative project whose theoretical sustainability effect, although perhaps lower, nevertheless, has fewer risks associated to it. The determination of an innovation project's risk structure can be oriented according to the following criteria.

Ecological Reversibility: When they in principal have an exit option, innovations should avoid "leaving tracks." Here, irreversible ecological damage is meant which remains after a technology, or market, no longer exists: accumulated emissions, resource inputs sealed-up surfaces, left-over waste, buildings, damaged or destroyed biotopes, loss of biodiversity, etc.

Conformity Flexibility: The correctability of a started development path is first of all a question of technical changeability. Here, design characteristics, such as e.g. a module building design or updateability, are meant. With the increasing non-material character of the innovation type (service, systems, organization, or institution), its conformity flexibility is not a technical, but a communicative issue. This includes cybernetic directional characteristics which are based upon social interactions. Participation models can open communication channels as a means to use stakeholder dialogue as an information deliverer or early warning system. Such mechanisms additionally allow a feedback between the operational innovation process and the public "opinion barometer."

Economic Reversibility: The economic reversibility of an innovation project can be increased in two ways. Supply-side "lock-ins" can be alleviated when investment is immobile and product-specific, i.e. irreversible capital is avoided. Demand-side "lock-ins" can be contained when the improvement substitutes for previously existing means and instruments meant for the fulfillment of needs, i.e. it ties down no additional routines and needs.

Avoidance of a High Infeed/Effective Power: When material-technical improvements or, changes are pending, preferable selections from the available ecological solutions are those which have proved to have the best environmental tolerances during the course of their co-evolutionary development history[4]. Additionally, the improvements displaying short space-time impact chains can be a revealed preference. Here, it is important to reduce the divide between

[4] Examples: Ecological farming; fishing rods made of pieced-together bamboo (instead of glass fiber plastic alloy or carbon fibers); shoes made of leather, linen, and natural rubber (instead of plastic); bicycle guard plates made of wood (instead of plastic, tin, or aluminum).

the range of human dealings in time and space on the one hand, and the knowledge of possible consequences of action on the other.

These criteria have consequences even for the selection of risk-reducing innovation types. New products and technical procedures display a direct ecological relevance due to their proximity to the physical-material realm, i.e. they necessarily induce or change material and energy flows. Contrastingly, new services, system solutions, as well as organizational and institutional changes have their starting point in the immaterial realm. Here, material effects reveal themselves only indirectly, notably through changed organizational structures, rules, and attitudes. Their chain of execution is, first of all, further-reaching in terms of cause orientation[5], and, second, more mouldable because the creation of ecologically relevant facts stands at the end of a long, particularly hierarchical causal chain. It is therefore no coincidence that modernization risks, which basically never represent anything more than innovation risks, are mostly dealt with under the category of "technology results assessment." The tendency therefore mostly leads towards product and technical innovations which are systematically connected to risk and rebound effects.

4 Timing in Innovation Processes

4.1 Configuration Boundaries and Decision Sequences

Taking the start of the implementation, i.e. the realization of a concrete innovation object as a time-based reference point, a sequential decision structure results with at least one "before" (*ex ante*) and one "after" (*ex post*) element. Therefore, all previous process phases can be classified into the area of *ex ante* control, whereby those phases dealing with the development of concrete innovation objects would rather be associated with a possible *ex post* control.

Within this time-based rough structure, the differentiation and exact progression of decisions are of great importance: The longer the concretization of the actual innovation object remains open, the greater the chance to react to any potential ecological or social detriments by means of corresponding adjustments. A process which passes through an "experimental phase" has the advantage, that it is more open to learning effects and interactions which can help the stabilization of the (sustainable) innovation direction. In other words: The innovation object should stay "mouldable" as long as possible in terms of having a high sustainability effect. A high level of formability can be achieved when the initial guidelines for the innovation are first limited to a

[5] Bierter [2], Schmidt-Bleek [18, pp. 67–70], and Stahel [20, p. 155] argue similarly, attributing greater efficiency advantages to new services and system solutions instead of product and technical innovations.

goal corridor which allows adequate room for experiment, adjustment, or optimization. Instead of a premature anticipation of the innovation type or the concrete innovation object, the initial problem alone could be formulated to be solved step-by-step over the course of the process. The decision-sequence to be tackled would offer the option to interactively shape each of the concretization steps, i.e. coupling it back to the external "guard rails." This results in a sequence which will be discussed in the following.

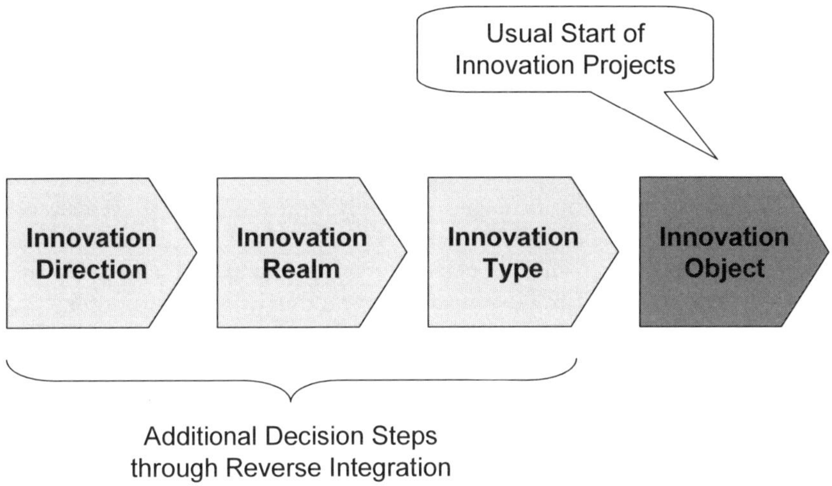

Fig. 2. Decision sequence through reverse integration

This sequential breakdown is identified as reverse integration. It explains how, based upon preliminary fundamental decisions, the "gates" for a certain innovation object are set. In principle, the above sequence can be reapplied from the start for each individual process. This would raise the possibility of routines, which predestine certain innovation types and objects, to be overcome. However, exactly this is normally not the case, as seen e.g. in the automotive industry. For nearly a century, and for the sake of satisfying mobility needs, this branch has offered only those solutions which comply to the type "product" and the object "car." As a result, the preliminary decision phase, which would answer the question of whether other kinds of innovation types (e.g. mobility services) or objects (e.g. bicycles, trams, and trains) would also be a possibility, was simply skipped or ignored completely. A high-tech automotive combine like Mercedes Benz therefore kicks off its innovation processes based upon a nearly 100-years-old method of making fundamental decisions. This is a sure way of blinding out causal and low-risk solutions from the start.

4.2 Reverse Integration

The way out of this strategic dead-end street is the reactivation of all preliminary decision realms, so that the rusty gates (staying with this metaphor) can once again be swung, and alternatives to the obdurate innovation directions can become possible. In particular, a management set on function orientation would have to take on a higher level of abstraction than the one dealing only with the drive and internal design of an already existing object called "car."

The four-level decision sequence shown in Fig. 2 relates to a "formation of circumstances according to their holistic capacity" [25, p. 78]. Through reverse integration, the realization of the innovation object is preceded by three decision steps, which can be seen as levels of an open hierarchical system [9] due to its increasing degree of abstraction. Therefore, more degrees of freedom develop, the higher the abstraction level, where the decision is made to the benefit of sustainable innovation. The cast-in-stone product and technical centrality found in many branches can only be overcome by reconstructing the "overlaying" decision levels. Here, not only could the request for an alternative use system to meet a certain need be made, but also possibly the general usefulness of entire business fields, or a particular business focus could be questioned.

The flip side of the coin recalls that an innovation process which ignores the highest hierarchy levels and starts directly on the level of innovation type "hardly has a chance to go beyond the horizon of previous" innovation practices. It will not reach the "corridors" which characterized the (previous) structures, organizational forms, or usage systems. It remains "tangled" in a web of innovation "routines" which almost never leave the technical dimension. Under these circumstances, only innovation prevents the vacating of an established technological paradigm and instead becomes an instrument of "structural conservatism" [19, p. 153].

This diagram joins the already mentioned conceptual elements and points out the following ideas:

- Increasing concretization – regardless of what stage of the process you are in – principally leads to a steady loss of formability and/or controllability of the innovation object. The target corridor (maintained by innovation management) in which the gradually developing innovation object moves along the timeline, opens itself in the shape of a funnel.
- The longer the process is "stretched out" due to the innovation decision being dismantled into successive concretization phases, the easier will be feedbacks from the external guard rails.
- The transition from the *ex ante* to the *ex post* control (perpendicular lines shown in Figure 3) additionally leads to a qualitative change. The start of the implementation phase is, at the same time, accompanied by the removal of uncertainty, thus, "perfect actualities" are created. A controllability is therefore only possible under certain conditions. The ideal case of an innovation path which could be controlled up to and including the

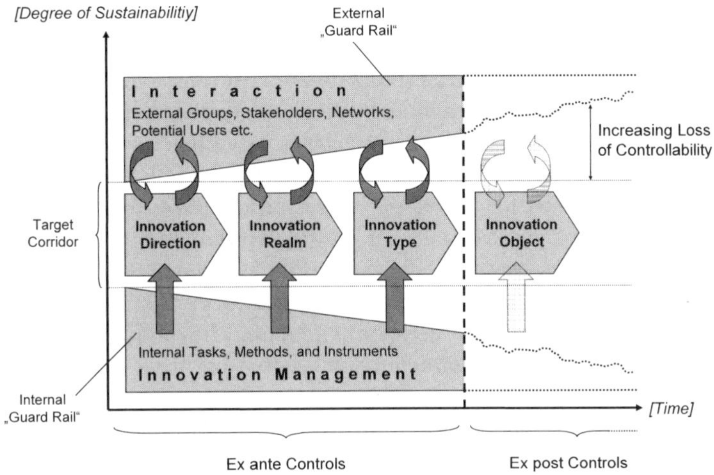

Fig. 3. Phases of an innovation process

point of market readiness is possible theoretically, but would hardly be encountered in the empirical world, and would also bring up the question of whether the object could still be named "innovation".

Nevertheless, the transition to *ex post* control is not to be interpreted as the "point of no return." Formation options still remain, even if only rudimentarily. Although systematically decreasing controllability forms the inescapable background of each innovation process, it can strongly vary based upon the respective risk characteristics of the innovation object. Consequently, controllability is less a question of "either or" than it is one of "more or less."

The design of the first three decision levels will, in the following, be assigned to *ex ante* control, and the content concretization of the innovation object, i.e. the fourth level, will be attributed to *ex post* control.

5 *Ex Ante* Control

5.1 Innovation Direction

The sustainability direction of a plan for innovation does not appear in a vacuum, but rather in close relation to history, competencies, general parameters, as well as additional specific attributes of the innovator. A strategic positioning by the company based upon any relevant market relationship is also included. The direction where the company can innovate to, without having to adjust its own assumption of disposition or overstep any core competencies, depends on the resulting path dependence. Only when this *ability* meets a specific *desire* a tendency towards sustainability-oriented innovation will result[6]. A normative management can deliver the basis for this desire, which Ulrich/Probst [21, p. 269f.] define as "development and implementation of a value system for the company [...], which is capable of establishing and legitimizing future company activities from a superior point of view, and creating a context with a point for all those involved and concerned." Borrowing from Ulrich [22] and Pfriem [14, p. 169ff.], normative management is represented as the highest of all management levels. The foundation of values and norms to be laid here should become effective in the form of orientation knowledge in all company-political activities, thus reaching beyond the "normative management" ⟶ "strategic management" ⟶ "operative management" chain of innovation management.

Along with such a goal-oriented or "offensive-minded" designation of a sustainability-oriented innovation direction, "defensive-minded" moments certainly also come into consideration. The starting point can be the search for solutions to a certain problem. Not only targeted search processes, but also a spontaneously occurring chance in the sense of "technology push" can be the initiator, assuming that the resulting goal direction pertains to "sustainability." Extrinsic impulses (market signals, new laws, Greenpeace "ante portas", etc.) can place external pressure on a company, resulting in an innovation project.

5.2 Innovation Realm

The determination of the innovation realm first implies the question of whether the innovation deals with an internal process, i.e. something concerning the inner workings of a company, or with a market-relevant innovation. In the former case, a determination of the respective company realm would be needed, while a determination of the relevant area of demand, function, or business field would be needed in the latter. For a chemical company desiring a more sustainable product line (innovation direction) in its "textile washing" realm, this could result in initially determining a relevant need, or function, e.g. "clean textiles without chlorine" or "clean textiles due to factor X being raised to increase resource productivity." Within this framework, product innovations would be only one of several possible solutions. An alternative would be the establishment of a new business field to meet such needs by an altered usage system in connection with the service of "clean clothes."[7]

[6] This does not mean that all sustainability innovations in a company must have goal-oriented processes as their result.

[7] For sustainable usage strategies in the laundry realm see Hirschl et al. (2001), pp. 54–57.

5.3 Innovation Type

A change in perspective is encouraged so that the process runs successively ("from above") and the previously determined innovation direction and the innovation realm converge with the implementation of the object (Figure 3). Thus, the next concretization step can be handled unbiasedly with regard to previous routines. From the viewpoint of satisfying a particular need or fulfilling a certain function, systematic and organizational innovations within a spectrum of possible innovation types are ranked equally with technical orientation or new products. Exactly the above mentioned ideas on ambivalence of the innovation principle are closely related to the need of giving such innovation types a high priority in the future whose effects are seen in new services, new usage systems, and a change in the consuming culture. System, organizational, and institutional innovations with low rebound and risk effects seem per se predestined for this.

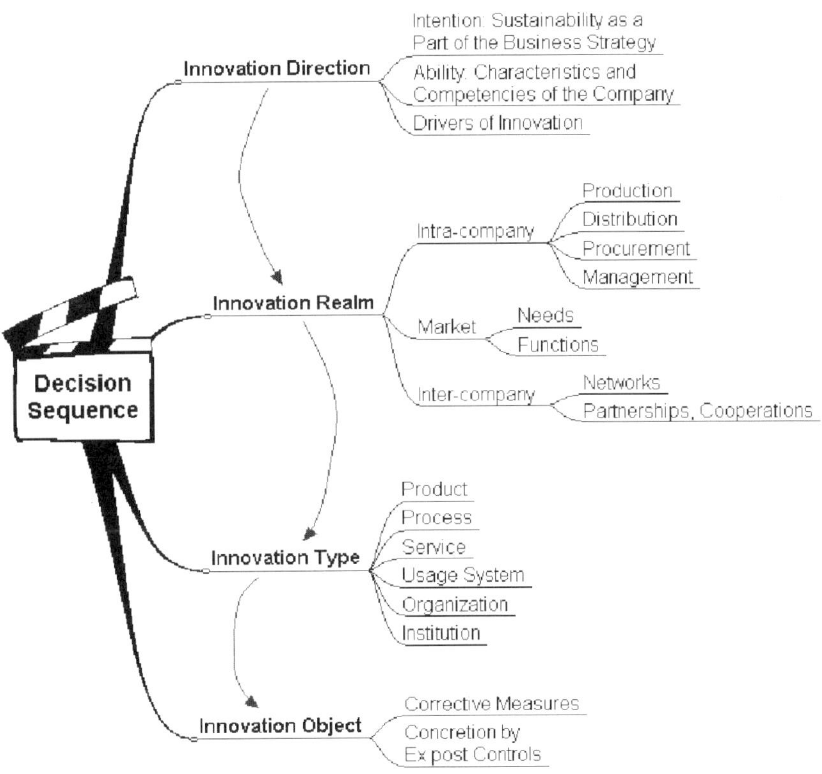

Fig. 4. Sequential innovation process

6 *Ex Post* Control

The decision for a certain innovation object marks the last phase before diffusion, and is characterized by a gradual resolution of uncertainty. Over the course of implementation, unforeseeable side effects are discovered on the one hand, from which an intervention/directional need may be derived if needed. On the other hand, this discovery occurs simultaneously with the creation of facts, thus greatly limiting the process controllability. Although this structural ambivalence is in principle unavoidable, starting points for control and directional functions can still be named. It depends on whether the content design within the realm of *ex ante* control was adequately oriented to the risk reduction criteria. Remaining tasks for *ex post* control include (among others):

- Control and monitoring,
- Acclimation, optimization, or substitution of technical and organizational details,
- Cancellation/termination of the project, should side effects occur over the course of market introduction for which an appropriate substitution is not possible,
- "Recalls",
- Flanking communicative measures which affect user behaviour and allow corresponding learning processes.[8]

The specification of the actual innovation objects also allows adaptability and optimization options for the purpose of increased sustainability effects. When e.g. a chemical producer has decided *ex ante* not only on the development of a new product (innovation type) in the demand field of "clean clothes" (innovation realm), but also on the innovation object "new laundry detergent", further options still remain *ex post* for the concrete design: Should it be a compact detergent? Should it be offered in powder, or liquid form? Should its use be according to the "building block" system? Which resources and inputs of what origin should be included into the production? How should it be packaged, and based upon what distribution system should it be marketed? What accompanying communication means (directions for use, etc.) can be used to optimize the reduction of its harmful side effects, etc.?

7 Forward Integration Through a Test Phase with Potential Users

Many side effects of an innovation are revealed during the course of trial, usage, or market introduction of the object, as the sustainability effect of an

[8] In a broader sense, these measures can also be understood as a stakeholder dialogue.

innovation is highly dependent on user behaviour. To determine the characteristics and routines of a "typical" user of the innovation object, it is possible to add an additional experimental phase to the innovation process. This could take place at the point between the *ex ante* and *ex post* stages and would attempt to create a usage context as close to the real world as possible. For this purpose, "test users" could be involved to reveal any possible negative sustainability effects, most notably:

- Technical,
- Performance-related, and
- Structural or overall system

effects. Particular attention should be paid here to rebound effects, whose occurrence is mainly connected to performance-related and structural aspects.

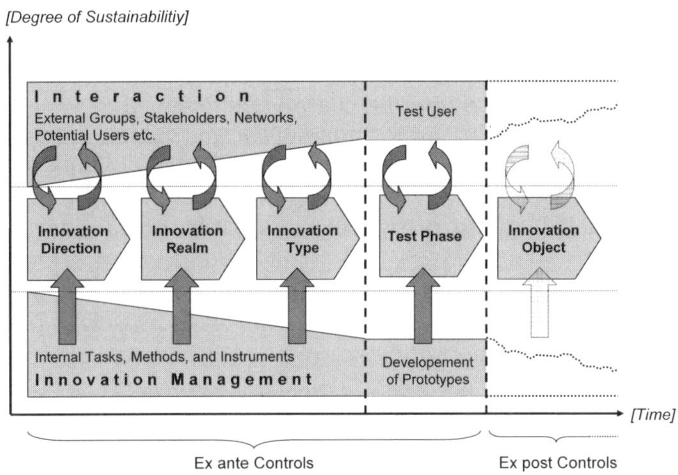

Fig. 5. Adding a test phase to the innovation process

This methodology draws from the "lead-user" approach developed by Hippel [7], but deviates from it in that the potential user must be suited in terms of an anticipation of sustainability effects which are first revealed during the use of the innovation object. Here (and not just out of necessity, as intended by Hippel) trend-setting, particularly interested, or creative users come into play who may turn out to be idea givers, or even inventors. Something just as sensible would be a (coincidental) selection of users who are not seen as "sustainable" lead users, but instead would be unbiasedly confronted with new solutions in the light of sustainability considerations. This kind of "forward integration" [8, p. 337] aims at better determining the behaviourally and culturally determined sustainability potentials and risks of an innovation.

8 Individual Provisions for Directional Certainty (Overview)

In closing, a few measures for increasing directional certainty should be discussed which, along with all previously mentioned phases, can also find application. These include instruments of knowledge management, particularly the acquisition of relevant information, data, scientific analysis, reports, expert opinions, etc., all of which can theoretically evaluate the risk structure of innovation objects being considered. Case studies, practical examples, documented projects of comparable plans as well as the identification of best practices also provide further in-depth information. The sharing of experts' experiences through networking, or cooperation with other companies can also provide additional input on risk determination. An increased usage of new communication media can improve control and monitoring, as the networking of the subsystems affected by the innovation and function areas is of great importance. Feedback which is as instant as possible provides an increase of directional security.

A timely embedding of the innovation process into a communicative exchange with relevant societal actors can increase the cognitive ability in terms of social and ecological damage potentials. Here, a correspondingly moderated "stakeholder dialogue" [3] can serve as a kind of an early warning system. In addition, risk generation and distribution requires a societal legitimization, otherwise conflict-laden negotiation processes may occur. The more timely socially affected groups can be integrated, the greater can be their initial influence "to reasonable consider (social) risks" within a discourse clarification. Each innovation, at the same time, means a transfer of both, chance and risks. An additional need for discourse is therefore displayed, namely for the purpose of clarifying how and to whom a certain, generally acceptable risk may be dispersed. Still, part of the responsibility is transferred from companies to society through participation by external groups. Through such interactive process creation, the innovator creates a double safeguard: First, the possibility of detriment is reduced, as the decentralized, sometimes implicit knowledge of external actors is used as a resource for directional certainty management. Second, the possibility of being the sole person or organization to blame in the case of loss or damage sinks.

Four different levels of social interaction between innovation management and external actors can be specified[9]:

[9] The sequence correlates with an increasing intensity of integration.

1. Coordination of directional certainty in a broader sense; external communication as an early warning system, with which potential dangers and damage can be diagnosed in a timely fashion, so as to countersteer if needed, e.g. information exchange through online and print media.
2. Reflexivity in terms of a discourse clarification about the meaning of sustainability by concrete implementation through an innovation process, for example active engagement of companies in sustainability and risk dialogues such as those found in "Agenda 21" projects.
3. Feedbacks with stakeholders, relevant actors, and societal subsystems as a means to legitimize specific innovations and the overall magnitude of risk transformation; installation of a "social early warning system": e.g. round tables or symposiums initiated and conducted by companies where critical interest groups and NGOs may participate.
4. Integration of lead users and other potential users as co-creators or co-producers of the innovation itself[10]: e.g. regular workshops with users and providers of ideas.

9 Conclusion: Risk Reduction as a Self-Contained Sustainability Principle

The ecologically and socially destructive power of modernization processes, which brought the topic of sustainability into being, is itself a result of previous innovations. Therefore, if it is to serve the purpose of sustainable development, a change of course cannot be achieved while the mode of change remains otherwise structurally unchanged. Even innovations meant to be sustainable can have unintended effects, leave scars, and accelerate growth in consumption. Sustainable strategies can thus sometimes also mean, when in doubt, stepping away from the "roulette wheel". However, this does not mean an outright denial of innovation as an important mode of change[11].

What is needed is an understanding of sustainability which, along with the actual ecological and social contents, also has process-related components of the necessary modes of change as its focus. Here, aspects like controllability, safety, and straightforwardness (in other words: risk reduction) would acquire self-contained goal attributes. Increased directional certainty of innovation processes does not just mean a minimalization of risks in the style of Seveso, or Chernobyl, but those of cumulative growth risks as well. The resulting consequences for operational management can be termed, according to [26], as "innovating innovation." Opening the way for innovation types which involve altered usage systems and behaviours, i.e. those that are more secure due to their immaterial character, depends to a high degree upon the sequential decision structure. A refinement of the classical instruments of innovation

[10] E.g. the "provider/user strategies" of Meyer-Krahmer/Jochem [11].

[11] Alternative modes of change would be e.g. exnovation, renovation, and imitation.

management is not sufficient here. The innovation process as such requires a different structurization, namely one not committed to the optimization of an innovation type which itself does not stand at disposition (mostly a product or production process). Its start should be returned to the decision level which the innovation direction, the innovation realm, and the various options for innovation types are located at (reverse integration). For starters, an additional degree of freedom would be achieved in terms of a potential solution to path dependencies. Also, the process would, in this way, finally become a *process*, namely one that is a sequential result of decision levels with increasing degrees of concretization. At each of these levels, risk reduction can be embedded as an additional selection criterion. Directional certainty then takes on a level of importance which extends beyond mere fine-tuning. Sustainability in innovation activities does not just mean managing existing risks better, but – through the choice of structurally safe alternatives (and when possible) – not letting them happen at all.

References

1. Beck U. (1992): Risk Society. Towards a New Modernity, London
2. Bierter W. (2002): System-Design: Radikale Produkt- und Prozessinnovationen. In: Jahrbuch Ökologie 2002, München: 171–193
3. Freeman R.E. (1984): Strategic Management: A stakeholder approach. Boston et al.
4. Gleich A.v. (1997): Innovationsfähigkeit und Richtungssicherheit. In: Gleich A.v., Leinkauf S. und Zundel S. (Hrsg.): Surfen auf der Modernisierungswelle? Ziele, Blockaden und Bedingungen ökologischer Innovation: 245–261
5. Gleich A.v. (1999): Vorsorgeprinzip. In: Bröchler S., Simonis G. und Sundermann K. (Hrsg.): Handbuch Technikfolgenabschätzung. Berlin: 287-293
6. Gronemeyer M. (2000): Immer wieder neu oder ewig das Gleiche – Innovationsfieber und Wiederholungswahn. Darmstadt
7. Hippel E.v. (1986): Lead Users: A Source of Novel Product Concepts. In: Management Science, 32. Jg: 791-805
8. Hübner H. (2002): Integratives Innovationsmanagement, Nachhaltigkeit als Herausforderung für ganzheitliche Erneuerungsprozesse. Berlin
9. Koestler A. (1967): The Ghost in the Machine. New York
10. Luhmann N. (1991): Soziologie des Risikos. Berlin, New York
11. Meyer-Krahmer F. and Jochem E. (1997): ökologische Innovationen aus technologischer Sicht. In: Gleich A.v., Leinkauf S. und Zundel S. (Hrsg.): Surfen auf der Modernisierungswelle? Ziele, Blockaden und Bedingungen ökologischer Innovation. Marburg: 71–92
12. Paech N. (2003): Innovationen und Nachhaltigkeit – Lösung oder Teil des Problems? In: Politische Ökologie, 21. Jg., Heft 84: 16–18
13. Paech N. and Pfriem R. (2002): Mit Nachhaltigkeitskonzepten zu neuen Ufern der Innovationen. In: UmweltWirtschaftsForum (uwf), 10. Jahrgang, Heft 3, September 2002: 12–17
14. Pfriem R. (1996): Unternehmenspolitik in sozialökologischen Perspektiven. Marburg

15. Pilardeaux B. (1999): Pythia, Medusa und Zyklop. In: Politische Ökologie, 17. Jg., Nr. 60: 25–26
16. Roller G (1999): Mehr schlecht als Recht. In: Politische Ökologie, 17. Jg., Nr. 60: 55–58
17. Sauer D. (1999): Perspektiven sozialwissenschaftlicher Innovationsforschung. In: Sauer D. and Lang C. (Hrsg.): Paradoxien der Innovation. Frankfurt, New York: 149–173
18. Schmidt-Bleek F. (2000): Das MIPS-Konzept. Weniger Naturverbrauch – mehr Lebensqualität durch Faktor 10. Mnchen
19. Simonis G. (1999): Die Zukunftsfähigkeit von Innovationen: das Z-Paradox. In: Sauer D. und Lang C. (Hrsg.): Paradoxien der Innovation. Frankfurt, New York: 149–173
20. Stahel W. (2001): Sustainability and Services. In: Charter M. and Tischner U. (eds.): Sustainable Solution. Sheffield: 151–164
21. Ulrich H. and Probst G. (1988): Anleitung zum ganzheitlichen Denken und Handeln – Ein Brevier für Führungskräfte. Bern, Stuttgart
22. Ulrich H. (1981): Managementphilosophie fr die Zukunft. Bern, Stuttgart
23. Weizsäcker C.F.v. (1977): Der Garten des Menschlichen. München, Wien
24. Weizsäcker C.F.v. und Weizsäcker E.U.v. (1984): Fehlerfreundlichkeit. In: Kornwachs K. (Hrsg.): Offenheit – Zeitlichkeit – Komplexität. Frankfurt, New York: 167–201
25. Wilber K. (1997): The Eye of Spirit – An integral vision for a world gone slightly mad. Boston
26. Wilde R. de (2001): Innovating Innovation. A contribution to the philosophy of the future [12]

[12] Keynote lecture at the 3rd POSTI International Conference, London, United Kingdom, 1-3 December, 2000;
`http://www.esst.uio.no/posti/workshops/dewilde.pdf`

Comment: Innovation Ability and Innovation Direction

Arnim von Gleich

University of Bremen, Faculty 4, Production Engineering, Technological Design and Development, Badgasteiner Str. 1, D-28359 Bremen.
Gleich@uni-bremen.de

The ways to sustainable economies are greatly dependent upon the ability of innovation. This is even true for nature conservation with its notedly "conservative" ecological goals. We can only effectively protect certain areas and ecosystems when just about everything "around and about" them changes. In this respect, it is completely understandable and relevant that the improvement of innovation ability is seen as a core element of all sustainability strategies. Innovation ability is a fundamental requirement of an efficiency strategy (see [8]). Potentials for improving resource efficiency are readily available in great quantities. The actual challenge is found in the ability of economic actors to recognize these potentials and (most of all) to make them a reality. Innovation ability is also the prerequisite of the strategy of consistency. The switch to renewable material and energy sources, and the embedding of an economic-technical "metabolism" into a natural one call for a rather far-reaching innovation ability (see [22, 23]). Furthermore, a sufficiency strategy with its far-reaching change in consumer behavior and lifestyle is massively dependant on the ability to restructure our life and economy. Consequently, all three strategies demand technical, organizational, institutional, and systematic innovations, all having various combinations. When the same people who, in the past, have fought against technicological risks like atomic energy, synthetic chemistry, or genetic engineering, and took to the streets against the ecological consequences of the exponential growth of material and energy consumption, now, for ecological and social reasons, join the campaign for an improvement in innovation ability, one has to ask why innovation ability and innovation actually hardly make any progress. When proponents of innovation and innovation ability can be found everywhere, then why do innovations still have it so hard?

The doubts regarding certain technological developments have certainly not vanished, and the often-scolded "doubters" have not fallen silent. Let us assume the case of an "overall general mobilization", that an all-out improvement of innovation ability would occur, that innovations would all succeed, started by incremental improvement innovations up to basic innovations up to far-reaching reforms "revolutionizing our life conditions". What would be the intended and the unintended effects of such a development? More sustainability? More prosperity? More growth? More resource consumption and environmental damage? Or more uncertainty and risks?

Is it possible to give improved innovation ability a direction? Can innovations be successfully navigated? Or to put it the other way around: Is a directionally *in*dependent improvement of innovation ability even imaginable? Are innovation ability and innovation direction two completely independent variables? If not, how closely are they connected? Is a form of innovation ability improvement conceivable and realizable (through e.g. the integration and support of certain actors in the innovation system) which at least raises the likelihood at the innovation moving in the direction of sustainability? Niko Paech deals with this question in the first part of his work "Innovation as an Ambivalent Mode of Change". The second part tackles the issue of how innovation risks, especially large risks, can be recognized and avoided in a timely fashion. Paech is rather sceptical about the possibilities of risk minimization. However, his contribution referring to the introduction of sustainability aspects through "proceduralization" through an extension and reorganization of innovation processes is certainly quite promising.

The following ideas are primarily enhancements of Paech's approach. They are a result of the experiences and insights from a research project concerning the replacement of hazardous substances with less dangerous materials and concentrating on the relationship between innovation ability and innovation direction with regard to the reduction of risks. Using 13 case studies, the research project "SubChem" examined the substitution of hazardous substances as an innovation process in companies and/or value-added chains[1].

1 Innovation Ability – Drivers and Restraints

In many well-known cases – for example the substitution of asbestos – the replacement of hazardous materials developed very slowly. Other hazardous materials, such as chromate in cement, methylene chloride in paint remover, or environmentally harmful or health-threatening heavy metals have, to this date, not yet been replaced. Due to their proximity, and in an attempt to understand this "tenacity" and persistence, the actors in the respective innovation systems are addressed first. The identification of "promoters" is then

[1] The project "Options for the Design of Innovation Systems for the Successful Substitution of Hazardous Substances" (SubChem) was funded by the Federal Ministry for Education and Research as part of the "General Framework for Innovations for a Sustainable Economy" program (FKZ 07RIW4). For further information, visit `http://www.subchem.de`, and see also [2].

aimed at the question who engages in promoting the process of change driven by concerns of environmental and consumer protection, quality assurance, or even sustainability itself. At the same time, the attempt is made to identify "blockers" who are motivated more or less by profit, indifference, and ignorance. In many cases, however, these blockers cannot be identified at all. Substitution was often not actively hindered; it simply made no progress, and instead bogged itself down.

1.1 System Inertness as the Greatest Innovation Restraint

According to this, the greatest innovation restraint is not found at the level of actors and their "motives". It has much more to do with "system quality," i.e. the "system inertness". Regardless of whether innovations are intended towards sustainability or some other direction, innovations are initially rather unlikely. In this respect, the concentration on innovation ability seems completely justified.

Still, innovations continue to occur. This means that there must be a driver which is capable of overcoming system inertness. This "system-moving" impetus is also absent at the level of actors and their direct "motives." The decisive factor has "system quality": market competition. In dynamic competition, companies are simply forced into innovations, even if they only want to maintain their existence. However, competition can only propel something forward; the ability towards innovation itself is highly dependent on general conditions, the architecture of the innovation system, the concrete actors, their interplay, and their chances for influence[2]. Those who cannot develop the corresponding abilities are excluded from the market. That is why it would not be sufficient to only intensify competition as a means towards "innovation promotion".

If system inertness is the strongest innovation restraint, and competition figures as the strongest means for overcoming this restraint, then innovation research needs to dedicate itself more to these systemic phenomena as the way towards opening design options[3]. Here, however, competition cannot only be characterised by its intensity, but most of all by its quality, the kind and type of market, and/or the type of competition. After all, the competitive situations of companies – and at the same time, no longer just the companies, but rather the entire value-added chains and innovation systems – are completely different in various markets.

[2] See [26, 30, 15, 7, 33, 21].

[3] System inertness is currently examined more closely than market quality and competition conditions. Current research on such topics as path dependencies, technological "lock-ins," investment cycles, lead markets, and "windows of opportunity" can be seen as aspects of research on innovation system inertness (see e.g. [29, 14, 10, 11, 19, 5, 17]).

1.2 Competition as the Most Important Driver

With the help of a rough "ideal type" differentiation between a "fordistic mass market" on the one hand (supply-dominant, often unsaturated, unified and stable with a predominant price competition and long product cycles), and a "differentiated quality market" on the other hand (demand dominant, saturated, fragmented, dynamic, with increasingly shorter product cycles, and many companies which pursue "trademark strategies"), the operation possibilities and restrictions of companies can be better understood. These competitive conditions constitute quite different realms of operation and susceptibilities of companies and/or value-added chains in light of external influences. Closely related to this are effects arising from a company's position within the value-added chain. For those manufacturers of a mass product operating in an extremely price-competitive environment with only loose connections to their customers (e.g. cement and concrete), even tiny changes in the cost structure can play a decisive role. On the other hand, auto manufacturers, who are close to their customers and are also known for operating under intense price competition, have long since removed certain hazardous substances from their products and production, although the customers are usually completely unaware of this and more or less indifferent to the material. The auto manufacturers point out (and rightfully so), that the end customer is usually not willing to pay a penny more for this extra effort, but the company substitutes nevertheless. In order to draw (positive) attention to themselves, to build trust in their products, and to win over customers, they follow a brand strategy and are therefore extremely susceptible to "scandals" in the media and general public. Companies which these competitive configurations apply to cannot afford such "thickheadedness" like the one found in the cement industry.

1.3 The Power of Scandalization

In view of the relevance of 'systemic' aspects, system inertness, and competition intensity and/or quality, the meaning of actors in innovation systems and their motives are put into clear perspective. In spite of this, in many case examples/studies one driver proved to be particularly strong: the role of the public, the media, and the civil society. Scandals are some of the most effective promoters of hazardous materials' substitution. In a few cases, public debate has not only influenced the economic realm, but to a certain extent government actors as well. This was especially true in the case of cleaning agents for metal surfaces. The substitution of chlorinated solvents took place, comparatively speaking, very quickly, although the field of metal cleaners is extremely complicated, complex, and not clearly laid out. Keeping innovation direction in mind, questions can furthermore be raised regarding the sense of some substitution measures. In this special case, the concept of "water-based systems" played an important role in terms of the substitution direction. The prevalent association of "water-based" with "environmentally friendly" gave the impression of a directional certainty which could, in fact, not hold up in a number of tests and inspections. In some areas of usage, the closed application of known hazardous substances would have been preferable to the half-open use of water-based systems whose problems were often underestimated and/or not yet fully known.

Thus, three elements can be debated which (can) give innovation processes a direction. The first is the understanding of "quality" in products and services, the second are the public and its scandal potential. Third, and in both instances, guiding principles play an important role. All three elements are closely interwoven. "Quality means that the customer comes back, not the product," is one way of elegantly putting it, or "Quality is defined by the needs of the customer." Product and service quality are determined in a complex (not always conscious, or even spoken) interaction process between suppliers with their notions of excellence on the one hand, and the customers with their notions of quality on the other. Still, the discourse on which of these notions of quality reveal themselves, keeping technologies, products, and services in mind, has become more general and more public. The "power of scandalization" revealed in the cases above is merely the aspect seen upon first inspection, but nevertheless part of a far greater development.

2 Innovation and Risk

The fundamental "opening of innovation systems" resulting from the intensification of competition, globalization, and market saturation, as well as customer incorporation and public involvement give direction to innovations. Thus the chances for sustainability innovations are greatly increased. Still, the extension of innovation systems and the involvement of "stakeholders" do not protect against massive errors, nor do they overcome the "ambivalence of innovations". This leads us to the second central question from Niko Paech's text. The example just shown regarding the transition to a "water-based system" in metal cleaning agents, assisted by the notion that water-based systems are *per se* healthy and environmentally friendlier, clearly displays this aspect. This may be bearable in the case above, and correctable without any further incident. Though, when dealing with technologies having a particularly great impact and socio-technical risks, the task of a precautionary "innovation assessment" takes on an entirely different dimension.

On the other hand, innovation and risk are inseparable. This in turn leads to the fact that remaining uncertainties, along with system inertness, are some of the greatest innovation hindrances. Along with this, in the case of conflict, many actors are more than happy to use the "kiss of death" argument of insufficient knowledge. Above all, people often take note of the many uncertainties associated with hazardous substance substitutions when changes to a previously well-established practice are planned. The innovator is expected to provide a greater, more disproportionate "load of evidence" than those who want to leave things the way they are. In this way, uncertainties are structurally adversarial to innovations.

2.1 Aspects of Innovation Assessment

The example of the worldwide elimination of the production and use of chlorofluorocarbons (CFCs) as part of the Montreal Protocol of 1987 can certainly be seen as an innovation success story. However, with the topic of innovation direction in mind, the CFC example is very thought-provoking. After all, CFCs were introduced in the 1930s (with the goal of risk reduction) as an intended safe substitute material for poisonous and/or flammable refrigerants and propellants, such as ether, ammonia, or methyl chloride. Nobody at the time could have known that a reduction in the risk of poisoning and explosion would be accompanied by a deterioration of the ozone layer and the numerous resulting effects of this process. Only nearly half a century later was the global risk of stratospheric ozone deterioration known and recognized. Even the substitution of the allegedly safer CFCs with other substances in the 1980s and 1990s cannot be seen as a complete success when keeping innovation direction, i.e. risk reduction aspects, in mind. With the hydrofluorocarbon tetrafluoroethylene (R134a), the substitute material currently used in most cases, the ozone problem is minimized, but as before, this gas with its long lifespan still contributes massively to the global greenhouse effect (see [9, 18]).

The introduction of CFCs as an allegedly safer substitute may represent an extreme case. Other examples – the replacement of asbestos with non-biodegradable mineral fibers in construction as well as the replacement of flammable hydrocarbons with non-flammable chlorinated hydrocarbons – show a serious orientation problem in terms of innovation direction. Such uncertainties additionally bring a halt to some substitution processes. A solution for the problem of uncertainty ostensibly seems to lie close to this matter, namely the simple demand for more insight and knowledge. Of course people will make efforts in exactly this direction to the best of their ability. Still, working towards more insight and knowledge does not really solve the problem. Instead, it very quickly hits relative and absolute boundaries. A better and more exact examination of materials in terms of their toxic and environmentally hazardous effects (a minimal data requirement) is meanwhile demanded as part of the European REACH system for the so-called "old substances" which are manufactured beyond a certain production amount. This, too, will take time, and will only involve a minimal data requirement.

Risk management can therefore not solely rely (not even in principle) on the knowledge of the foreseeable effects of innovations. It is also not possible to wait with innovations until all possible, or expected effects are known and/or determined. The fact that something is "new" (and, as a result, still has unknown possible effects) cannot be a sufficient justification for far-reaching measures according to the precaution principle. Therefore, an appropriate handling of not knowing and uncertainties, two things which are always associated with innovations (see [3, 4]), is necessary.

2.2 Dealing with Uncertainty and Not Knowing – An Attempt to Operationalize the Precautionary Principle

Due to insufficient awareness of possible effects, we must initially revert back to trial-and-error strategies. This does not have to be the worst choice. A responsible trial-and-error strategy which adheres to the precautionary principle nevertheless requires a rational framing of the search domain. It holds its line where "too much at one time" is risked, i.e. where in the worst case scenario an error in the course of one trial would cause global and/or irreversible negative outcomes. This has been the case with extremely potent technologies such as nuclear power with its eternally radiating waste, as well as the release of self-replicating, genetically manipulated organisms, because too much can "happen" in the case of a delay in error detection.

Still, a slowly moving "trial-and-error" strategy must also be socially organized. Experiments should be done in such a qualitative and quantitative way which leaves an exit option to return back to the start or repair things in the case of a "trial" going completely wrong. This means that "learning" and "experiment" realms need to be created consciously with clearly determined boundaries according to technical as well as to economic risk aspects. Small steps and a slow increase in amounts should (when possible) be preferred, accompanied by an intensive monitoring of the consequences which become evident.

2.3 Agent Characterization (Hazard Characterization)

More and more purposeful things are, of course, possible beyond trial and error when rationally dealing with uncertainty and not knowing. An important access to this exists in the "characterization" of substances and/or technologies. The estimation of effects is essentially reliant on knowledge from three realms: knowledge of the triggering "agent" (material, technology, product), knowledge about the target system in question (application realm, exposed organism and/or ecosystem), and finally, the scientifically establishable effect model which can help explain how exactly the agent impacts on the target system (e.g. causing cancer, damaging reproductive organs, damaging the climate, etc.). When the target systems as well as the effect model are still unknown, the characterization of the agent nevertheless remains a starting point for operationalizing the precautionary principle (see Fig. 1).

The characterization of the agent (known as "hazard characterization" in toxicology) can independently deliver clear indications regarding the expectable width of steps (depth of intervention) and effect spectrums. Thus, the release of chemicals having certain (bio-) physical characteristics (e.g. persistent as well as water- and fat-soluble and mobile in various environmental

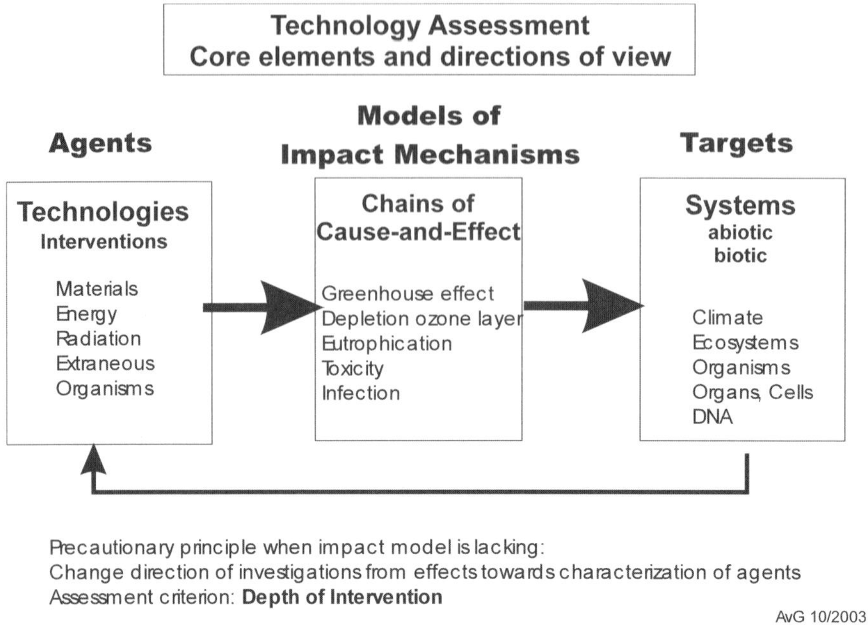

Fig. 1. Technology assessment – core elements and directions of view

media and, as the case may be, still bioaccumulative) contradicts the request for smaller steps and/or reversibility (see [32]). The EU chemical regulation already takes this into consideration. Very persistent and very bioaccumulative substances are, in principle, subject to authorization, even when no scientifically proved suspicion of problematic effects can be found. In addition, due to an ever-increasing understanding of various effect mechanisms on the molecular level, it is possible to obtain general indications of molecular configurations' estimable effect spectrum solely based upon observation (QSAR) (see [24]).

This knowledge can be applied not only for result estimation, but in a far more focused sense for the development and creation of materials and technologies. After all, materials and technologies do not simply appear out of thin air. They are developed and created by actors in innovation systems. The implementation of the precautionary principle therefore does not have to limit itself to "finished" chemicals, technologies, or products. The goal of sustainability, and especially the often-neglected goals of health and environmental friendliness, can and should be incorporated into their development and creation.

3 Two Approaches Towards a Sustainability-Oriented Design of Innovations

The focus of these considerations concerns a strongly expanded understanding of risk management which takes a foothold right from the start of development and creation of innovations. Two approaches towards action appear particularly promising: on the one hand, goal-oriented development and creation of materials and/or chemicals, technologies, and usage systems (following e.g. the goal of intrinsic safety), and on the other hand, the integration of the request and demand for worker, consumer, and environmental safety in company and, most of all, intercompany (value-added chain-related) quality management.

3.1 Design of Technologies, Processes, and Products Following Guiding Principles

As seen from the results of innovation and technology genesis research, guiding principles ("Leitbilder") play an important part in the creation of materials, technologies, processes, and products[4]. Those wishing to influence and to form innovations with the help of guiding principles must try to understand the effect requirements and effect nature of successful models. Such principles exercise their impact by motivating, constituting a group identity, coordinating and synchronizing the activities of individual actors, reducing complexity, and structurizing perception. Some of the most important requirements for the effectiveness of these principles are therefore their pictorial quality and emotionality, their orienting function, as well as their relation to wishes and feasibility. In other words: their ability to resonate in the consciousness of the actors[5]. The starting points for a concretization should be obvious. Thus, "doing business sustainably" is a goal which is too complex, too abstract, and too defensive. Demand-related models as e.g. "Sustainable Housing and Living" developed by the Enquête Commission of the German Bundestag *Protection of Mankind and the Environment*, or goals found on the intermediary concretization and operationalization levels, such as the concept of the "closed-loop economy," bionics (modelled after nature), and "green chemistry", or "sustainable chemistry" (see [16, 1]) would be more effective.

3.2 Comprehensive Quality Management of Value-Added Chains

In increasingly complex dynamic systems (not least of all in view of ever-shorter product cycles), the orientation is getting increasingly difficult for the suppliers as well as for the consumers. Here, risk communication along the value-added chain increases in importance. Transaction costs rise for everyone involved. Building trust would open a way towards reducing complexity and lowering transaction costs. Trademark strategies aim at this goal.

[4] See [6, 12, 13, 20, 27, 28].

[5] See [12, 13, 27].

Still, trust is something which must be earned. Legitimation by transparency, through traceability of decisions, and through integration of the precautionary principle (in general, legitimation through procedures) are possible means for improvement.

The attempt to address the issues of worker, consumer, and environment protection has so far mostly been done for separate business management systems. However, for a fairly long time now, the tendency has been towards an integrated management which simultaneously considers a variety of these aspects. Such an enhanced and integrated quality management must increasingly integrate into the entire value-added chain. This is especially important for companies and/or value-added chains which are particularly vulnerable to the threat of public scandal. For an increasing number of companies in industrialized countries, market chances can only be found in competition through *quality*, due to pressure from countries which can produce at a "cut rate" thanks to lower wages. However, for this purpose, and for building the necessary customer trust and loyalty for acquiring access to the "premium segment" found in almost all markets, considerable quality assurance investments are necessary. The avoidance of image-damaging scandals, lawsuits, or recalls has certainly become an important impetus for activities aimed in this direction.

4 Conclusion

Prominent phenomena like globalization and market saturation in industrialized nations (combined with increased competition and a switch from supply- to demand-dominated markets), both of which are intensified by trends like invidualization (due to the dynamization and fragmentation of markets as well as the shortening of product cycles) as well as the strengthening of civil and private society, media, and public in general, have all brought about an "opening of innovation systems" in modern industrialized society. The "susceptibility" of companies has increased dramatically. Businesses today intent to create customer loyalty by means of brand strategies. They attempt to "earn" the trust of customers and investors through management systems and long-term precautionary strategies (legitimation through procedures). Observation indicates that this "opening of innovation systems" results in a closer interrelation between innovation ability and innovation direction to the extent that (at least in certain markets) a directionally independent improvement of innovation ability does not (no longer?) exist.

There is no doubt that companies have to make profit as simple economic logic demands. Still, it remains unclear how businesses can successfully earn "sustainable" profits under the current conditions of competition. The economic future of companies may, to a great extent, rely on quality production and quality competition. A pure price–and–cost competition with developing countries is unwinnable. The direction and "quality" of innovations is more or less determined by explicit interaction between suppliers and customers, but is also discussed in the public realm. The public discourse refers to more than just "power of scandalization." In the future, skillful arrangements will need to limit innovation risks, the widths of the development steps, and the depth of intervention on the one hand. On the other hand, it is important to develop an understanding of how much is actually "on the line." Risk-free innovations are simply not possible. Finally, the public discourse on risks and scandals must be expanded by means of a "powerful" discussion on positive models and guiding principles, which in turn can be similarly effective in creating the demand for certain quality standards.

References

1. Ahrens A. und Gleich A.v. (2002): Von der Kreislaufwirtschaft zur Nachhaltigen Chemie – Leitbilder in der Chemikalienentwicklung und Stoffpolitik. www.subchem.de/startgerman.html
2. Ahrens A., Braun A., Effinger A., von Gleich A., Heitmann K. and Lißner L. (2005): Hazardous Chemicals in Products and Processes – Substitution as an Innovative Process. Physica Verlag, Heidelberg
3. Beck U. (1996): Wissen oder Nicht–Wissen? Zwei Perspektiven "reflexiver Modernisierung". In: Beck U., Giddens A. und Lash S.: Reflexive Modernisierung. Eine Kontroverse. Frankfurt a.M.
4. Beck U. und Bonss W. (2001) (Hrsg.): Die Modernisierung der Moderne. Suhrkamp, Frankfurt a. M.
5. Beise M. and Rennings K. (2003): Lead Markets of Environmental Innovations: A Framework for Innovation and Environmental Economics. ZEW Discussion Paper No. 03–01, Mannheim
6. Bijker W., Hughes T.P. and Pinch T. (1987)(eds.): The Social Construction of Technological Systems. New Directions in the Sociology and History of Technology. MIT Press, Cambridge, MA
7. Blättel-Mink B. und Renn O. (1997) (Hrsg.): Zwischen Akteur und System. Die Organisation von Innovation. Opladen
8. Bleischwitz R. (1998): Ressourcenproduktivität – Innovationen für Umwelt und Beschäftigung. Springer-Verlag, Berlin, Heidelberg
9. Böschen S. (2000): Risikogenese: Prozesse gesellschaftlicher Gefahrenwahrnehmung: FCKW, DDT, Dioxin und Ökologische Chemie. Leske + Budrich, Opladen
10. David P.D. (1985): Klio and the economics of QWERTY. In: American Economic Review. Papers and Proceedings 75/1985: 332–337
11. David P.D. (2000): Path dependence, its critics and the quest for "historical economics". Working Paper, All Souls College, Oxford & Stanford University, June 2000, http://www-econ.stanford.edu/faculty/workp/swp00011.pdf
12. Dierkes M., Hoffmann U. und Marz L. (1992): Leitbild und Technik – Zur Entstehung und Steuerung technischer Innovationen. edition sigma, Berlin
13. Dierkes M. (1997) (ed.): Technikgenese. Befunde aus einem Forschungsprogramm. edition sigma, Berlin

14. Dosi G. (1982): Technological paradigms and technological trajectories. In: Research Policy 11/1982: 147–162
15. Edquist C. (1997) (ed.): Systems of Innovation – Technologies, Institutions, and Organizations. Printer Publishers, London, Washington
16. Enquête-Kommission des Deutschen Bundestages Schutz des Menschen und der Umwelt (1997) (ed.): Zwischenbericht – Konzept Nachhaltigkeit, Bonn, (The Concept of Sustainability – Prerequisites for Tomorrow's Society – Abridged English Version of the Interim Report, Bonn)
17. Erdmann G. (1999): Zeitfenster beachten. Möglichkeiten der Ökologisierung der regulären Innovationstätigkeit. In: Ökologisches Wirtschaften 2/1999: 21–22
18. European Environment Agency (2002): Late lessons from early warnings: the precautionary principle 1896-2000. In: environmental issue report No 22, `http://reports.eea.eu.int/environmental_issue_report_2001_22/en`
19. Freeman C. and Perez C. (1988): Structural crises of adjustment: business cycles and investment behavior. In: Dosi G., Freeman C., Nelson R., Silverberg G. and Soete L. (eds.): Technical Change and Economic Theory. London, Pinter Publishers: 38–66
20. Hellige H.D. (1996): Technikleitbilder als Analyse-, Bewertungs- und Steuerungsinstrumente: Eine Bestandsaufnahme aus informatik- und computerhistorischer Sicht. In: Hellige H.D. (ed.): Technikleitbilder auf dem Prüfstand. Leitbild-Assessment aus Sicht der Informatik- und Computergeschichte. edition sigma, Berlin
21. Hemmelskamp J. (1999): Umweltpolitik und technischer Fortschritt. Physica Verlag, Heidelberg
22. Huber J. (2001): Ökologische Konsistenz. Zur Erläuterung und kommunikativen Verbreitung eines umweltinnovativen Ansatzes. In: Umweltbundesamt (ed.): Perspektiven für die Verankerung des Nachhaltigkeitsleitbildes in der Umweltkommunikation. UBA-Berichte 4/01, Erich Schmidt Verlag, Berlin: 80–100
23. Huber J. (2004): New technologies and environmental innovation. Edward Elgar, Cheltenham
24. Jastorff B., Störmann J. und Wölcke U. (2003): Struktur-Wirkungs-Denken in der Chemie – Eine Chance für mehr Nachhaltigkeit. Universitätsverlag Aschenbeck&Isensee, Bremen, Oldenburg
25. Lau C. und Böschen S. (2001): Möglichkeiten und Grenzen der Wissenschaftsfolgenabschätzung. In: Beck U. und Bonss W. (Hrsg.): Die Modernisierung der Moderne. Suhrkamp, Frankfurt a.M.
26. Lundvall B.-A. (1992): User-Producer Relationships, National Systems of Innovation and Internationalisation. In: Bengt-Åke Lundvall (ed.): National Systems of Innovation. Towards a Theory of Innovation and Interactive Learning. Pinter Publisher, London
27. Mambrey P., Paetau M. und Tepper A. (1995): Technikentwicklung durch Leitbilder. Neue Steuerungs- und Bewertungsinstrumente. Frankfurt a.M.
28. Meyer-Krahmer F. (1997): Umweltverträgliches Wirtschaften. Neue industrielle Leitbilder, Grenzen und Konflikte. In: Blättel-Mink B. und Renn O. (Hrsg.): Zwischen Akteur und System. Die Organisation von Innovation, Opladen
29. Nelson R. R. and Winter S. G. (1982): An Evolutionary Theory of Economic Change. Cambridge, MA, London
30. Nelson R. R. (1993) (ed.): National Innovation Systems: A comparative study. Oxford University Press, Oxford and New York

31. Paech N. (2005): Nachhaltige Innovationen: Zur Gestaltung ambivalenter Prozesse des Wandels, in: Jahrbuch Ökologische Ökonomik, Marburg, 225-250.
32. Scheringer M. (2002): Persistence and Spatial Range of Environmental Chemicals: New Ethical and Scientific Concepts for Risk Assessment. Wiley-VCH, Weinheim
33. Weyer J. et al. (1997): Technik, die Gesellschaft schafft – Soziale Netzwerke als Ort der Technikgenese. edition sigma, Berlin

Part III

Arrangements in Society and Economy Towards Sustainability

Deceleration – Revealed Preference in Society and Win-Win-Strategy for Sustainable Management. Concepts and Experimental Evidence

Edeltraud Günther[1] and Marco Lehmann-Waffenschmidt[2]

[1] Technical University of Dresden, Department of Economics and Business Administration, Environmental Management, D-01062 Dresden. Bu@mailbox.tu-dresden.de
[2] Technical University of Dresden, Department of Economics and Business Administration, Managerial Economics, D-01062 Dresden. manaecon@mailbox.tu-dresden.de

1 Introduction

Until recently 'deceleration' has been little recognized as a technical term, or as an idea. However, it seems to be getting more attention now. For example, the German magazine STERN dedicated in 2005 a cover story to deceleration, in the Anglo-American world, the "Quiet Life Hypothesis" is gaining followers, the "Heidelberger Club für Wirtschaft und Kultur" ("Heidelberg Club of Economy and Culture") dedicated its annual meeting in 1998 to deceleration,[3] and the competition for the German Study Award of the Körber Foundation in 2002 had the motto "Speed – the accelerated world."[4] In Italy, you can even study "Slow Food", and along German motorways you find signs with the slogan "be relaxed – just discover."[5]

[3] Heidelberger Club für Kultur und Wirtschaft (ed.) (1999): Im Rausch der Geschwindigkeit, Springer Verlag. To be sure this title meaning "the rapture of speed" should be understood in a critical, not an affirmative manner.

[4] "Tempo! – die beschleunigte Welt", forschen – Das Magazin des deutschen Studienpreises, Heft 1, 2003.

[5] Cf. also the report in "Die Zeit", 28.12.2006, "Dossier: Auf der Suche nach der verlorenen Zeit", S. 13-15, and the webpage www.zeit.de/2007/01/zeit. In fall and winter 2005/2006 the authors of the present contribution organized in cooperation with the "TU-Umweltinitiative" of the Dresden University of Technology an interdisciplinary series of lectures ("Umwelt-Ringvorlesung") on the subject: "Tempo! Tempo? – Beschleunigung und Entschleunigung im interdisziplinären Spannungsfeld" ("Speed! Speed? - Acceleration and Deceleration in the interdisciplinary area") (see:

Without any doubt, time is a decisive factor for the productivity and competitive advantages of companies. Still, more speed by continual, or even accelerated, acceleration may well be counter-productive and lead to an "acceleration paradox" – for example by product life cycles which are too short and therefore increase the share of R&D costs or by "Pyrrhus" victories which lead to "the winner's curse" instead of a stable market position. This acceleration paradox may show up in consumption, too. Consuming requires time and therefore competitors not only fight for their share of the consumers' cost budget, but also for their share of the consumers' time budget. It is this time budget which must be split up into productive, consumptive, and all other leisure activities, such as going for a walk or playing chess, which are neither productive nor consumptive in an economic sense. The wide range of consumption goods and the increase in consumed goods and services together with the already mentioned shorter life cycles, e.g. of computers, cell phones, or electronic equipment, are perceived by the consumers more and more as acceleration and personal burden. Speed can threaten the "happiness" of the consumers, and so acceleration may become an "acceleration trap" for business and society[6].

The term "deceleration" seems to be adequate for describing the opposite of acceleration. However, is there truly a preference for deceleration in the society, and can deceleration become a paradigm in business management? These questions give the impulse for the research presented here by asking four questions: What are the reasons for acceleration in business and society? What have been the consequences of acceleration so far? Can deceleration contribute to sustainable management? Is there a preference for deceleration in society, and how can it be measured?

2 Reasons for and Development of Acceleration in Business and Society

In this Section we will describe three levels of the emergence and spread of the acceleration phenomenon: on the macroeconomic, the microeconomic, and the motivational and behavioural levels.

`http://tu-dresden.de/die_tu_dresden/fakultaeten/fakultaet_wirtschaftswissenschaften/vwl/me/forschung/projekte/abgeP`. This series of lectures followed a students' seminar on the same subject in summer and fall 2005. The seminar papers can be found in: Guenther, E./ Lehmann-Waffenschmidt, M. (Hrsg.): Entschleunigung von Konsum- und Unternehmensprozessen, Dresdner Beiträge zur Lehre der Betrieblichen Umweltökonomie, Nr. 20/2006, `http://hsss.slub-dresden.de/documents/1157450611775-7808/1157450611-775-7808.pdf`

[6] Traps and treadmills jeopardising the happiness of modern mankind in developed countries are analysed by M. Binswanger: "Die Tretmühlen des Glücks", Herder Verlag, Freiburg 2006.

2.1 The Macroeconomic Level

From the macroeconomic perspective acceleration is familiar: Economic growth, reflected in a constant rate of growth and the resulting exponential growth curve, expresses acceleration. While modern economic systems aim for growth, they equally aim for acceleration. The reasons for growth and acceleration, which have been discussed for years now, are multiple. The range of reasons reaches from the institutional conditions of economics, such as the compound interest and employment problems due to technologically caused productivity growth, to psychological aspects of an elementary need of modern human beings to be equal to God[7].

Yet, do economies really grow exponentially? Analyzing the real development since World War II, for all developed countries – some exceptions omitted – no exponential growth of the aggregated economic performance but rather a linear trend can be shown. However, at least partially the money supply grew exponentially due to compound interest. The dynamics caused by this misalliance can lead to a misbalance for the developed countries which may even threaten their wealth. Still, beside this inherent explosive force of our economic system based on endogenously produced credit money, there is another threat from exponential and also linear economic growth – the overuse of natural resources. Section 3.1 will describe these threats in more details.

2.2 The Microeconomic Level of the Company

From a company's perspective, the reasons for acceleration can be identified if the question, who determines the handling of time in companies, is determined. Therefore three sources can be identified: The consumers and the environment as stakeholders in the handling of time, and the companies themselves through being affected by these stakes and by reacting to them one way or another.

The consumers set "point-of-time requirements" by requiring delivery at a specific target moment. This may be expressed by the characteristics timeliness (delivery at a fixed point of time, e.g. just in time), recentness (regarding existing conditions, such as legislation), and novelty (respecting new developments, such as the use of / emergence of renewable energies). Recentness and novelty may be in rivalry, as existing legislation may block new technologies, e.g. "genetic engineering". Moreover, the consumers set "period-of-time" requirements by requiring delivery within a certain time frame. Reasons may be expected time savings (e.g. maintenance within 24 hours), or flexibility (e.g. independence of office hours by internet banking).

[7] Cf. e.g. Lehmann-Waffenschmidt's contributions "Geld, Wirtschaftswachstum und Glück", in: "Wege in den Postkapitalismus", Hrsg. K. Woltron, H. Knoflacher, A. Rosik-Kölbl, S. 144 - 184, edition selene, 2004, and "Vision und Kritik der modernen Wirtschaft in Goethes 'Faust"', in: "Faust-Jahrbuch", Band I, Hrsg. B. Mahl, T. Loerke, Francke Verlag, S. 69 - 112, 2005.

The environment sets restrictions in three ways, which may reduce the choice set for companies:

1. the rate of reproduction (defined as 1 / time period of a complete renewal of resources in years) as a measure for the supply function of the environment with renewable and non-renewable resources,
2. the rate of decomposition (defined as 1 / time period of a complete decomposition of emissions; half times describe the rate of decomposition for exponential decomposition processes) as a measure of the carrier function of environment for conducts, i.e. non-desired output, such as "sewage", waste, and polluted air,
3. the rate of regeneration (defined as 1 / time period of a reconstitution to the original state) as a measure for the regulation function of the environment which interlinks the supply and the carrier function.

Embedded in these requirements of the consumers and the environment, the companies have to find the proper measure of time, that is, they have to optimize their time target. So far, however, the answer has usually been to increase the speed of their processes, because acceleration allowed time-dependent demands (timeliness, recentness, novelty, time savings, and flexibility) to be satisfied, thus creating competitive advantages ending in price premiums. As market cycles are restricted, the first supplier on a market (pioneer) can completely capture the market, whereas the follower, whose R&D time is longer, can only capture a reduced market volume, thus having to make profit sacrifices. Moreover, time strategies open up potentials for cost reduction.[8] For example, throughput times can be shortened by a change in production and stock, thereby reducing the capital employed.

2.3 The Level of Human Motivation

It is part of economic thinking to ask for the deeper motivation of consumers for acceleration, even if this question requires knowledge of other disciplines, such as psychology, or anthropology. Before consumption becomes a burden for people, there seems to be a long period, which our society has not yet passed, where acceleration in consumption is perceived positively.[9] Leaving aside that perception is intentionally influenced by the mass media, the question remains: Where does the consumers' motivation and willingness for accelerated consumption come from?

Modern research answers this question with psychological arguments. So, G. Scherhorn sees, like E. Fromm ("Haben oder Sein" — "To Have, or to Be") or H.E. Richter ("Der Gotteskomplex" — "The God Complex"), an elementary need of modern human beings to become like God ("Entgrenzungs- und

[8] Baum H.-G., Coenenberg A.G. und Günther T. (1999): Strategisches Controlling. 2. Aufl., Stuttgart: 154–161.

[9] cf. e.g. Gross P. (1998): Die Multioptionsgesellschaft. Frankfurt: Suhrkamp Verlag.

Gottgleichheitsbedürfnis" — Desire for Delimitation and Equality with God) by overcoming the essential human limits. This can be the hidden engine for the modern Consumers' behavior. Simply stated: The fear of loss (e.g. loss of security in religious or feudal societies or mortality) is overcompensated by human activities which realize the similarity, or even equality with god as promised in the Old Testament and other early Jewish and Christian texts. Consumption is a platform for realizing this "salvation", as permanently accelerated consumption gives the illusion of infinite determination by humans who perceive themselves as the creators of their own world[10].

3 The Consequences of Acceleration

In Section 2 the reasons and the development of acceleration in business and society were presented, and some of the consequences were already shown. These will be elaborated in more details in this Chapter.

3.1 The Macroeconomic Growth-Related Illusion of Acceleration: The Acceleration Trap

In the late 1990s there was much discussion between Herman Daly and other critics of growth on the one side, and the Nobel prize winner Robert Solow and other advocates of growth on the other. Neoclassical theory shows a remarkable, substantial contradiction in the heart of its theory: Neoclassical theory is based on self-restriction by negative feedback and by the definition of optima and balances — for the theory of growth, however, neither one is true. Instead of an optimum or a balance of the analyzed variables in absolute terms, the theory of growth defines optimal rates of growth and hence postulates an exponential, infinite growth of the considered variables in absolute terms. However, at the same point the potential infinite growth of physical economic variables meets the limits of the physical resources. Therefore, the belief in growth must be an illusion, unless technical progress and the dematerialization of consumption and production allow an infinite, sustainable economic growth based in value, not in physical terms.

This is the focus of the recent discussion of "weak" vs. "strong" sustainability between the critics and the advocates of growth. Can the speed of linear growth – or even an accelerated speed of exponential growth – be maintained sustainably without endangering the natural resources in a way that economic artifacts, such capital goods, consumption possibilities, and institutions can no longer regenerate them? Or does the belief in economic growth inducing a limitless wealth increase become a growth illusion and trap?

[10] See e.g. Lehmann-Waffenschmidt "Geld, Wirtschaftswachstum und Glück. Das Psychogramm unserer Zeit in Goethes 'Faust"', in: "Geld regiert die Welt", A. Karmann, J. Klose (Hrsg.), S. 285 - 308, Metropolis Verlag, 2006.

3.2 The Microeconomic Company-Related Illusion of Acceleration: The Productivity Trap

Even if only economic aspects are taken into consideration, phenomena such as the acceleration trap, show that it may be senseless to accelerate processes limitlessly i.e. that there are limits of acceleration.[11]

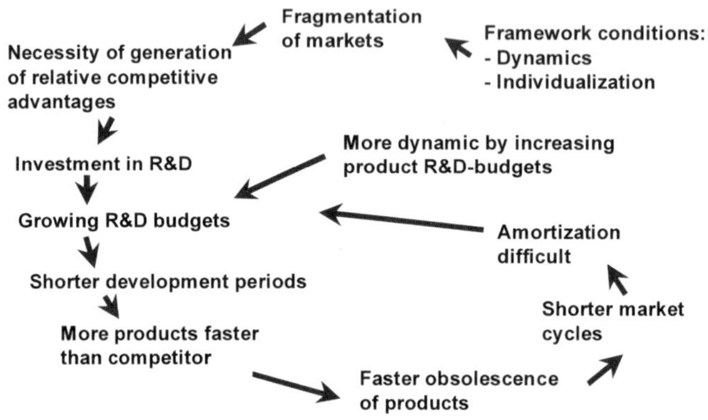

Fig. 1. Mechanism of the acceleration trap

The starting point for this mechanism are the framework conditions which can be characterized by a dynamic development – related to competition – and by individualization – related to the customers. The consumers ask for products which are adopted individually to their existing, or created needs. The companies try to avoid price and cost competition by differentiating their product range. This leads to a fragmentation of markets. For a firm to distinguish itself from its competitors, it is necessary to create many different relative competitive advantages. Therefore, extensive investments in research and development are necessary. Hence, the budgets have to increase annually. Consequently, the development periods decrease, so that the company can enter the market with more products in a shorter period of time. This also means that the existing products become obsolete faster, i.e. they have to become outdated to create demand for the new products. Overall, the market cycles become shorter, and amortization becomes more difficult. If the reaction is to increase the R&D budget to become even faster, the circle is repeated and a dynamic, self-enforcing process is started. If there is only one acceleration, a bigger portion of the market volume can be captured ("flash in the pan"). If there is a continuous acceleration, the sales decrease due to the shorter

[11] Cf. von Braun C.-F. (1991a): Die Beschleunigungsfalle. In: Zeitschrift für Planung, 2. Jg., 1991, Heft 1: 58ff.

market cycles. This effect is called an "acceleration-resistant sales-slide"[12] by Backhaus. Empirically, von Braun shows this acceleration trap for American companies[13].

From the ecological point of view the acceleration of processes shows consequences if time measures are not respected, as nature sets restrictions. These consequences refer to the already mentioned functions of the environment, the supply function ("the source runs dry"), the carrier function ("the valley is filled"), and the regeneration function ("the channel is blocked"). They can be analyzed with respect to two types of scarcity: the scarcity of rate and the scarcity of accumulation. The scarcity of rate asks for a critical rate of extraction (e.g. for renewable resources), of carrying capacity (e.g. of air), or of regeneration (e.g. water). The environment can tolerate a critical rate where self-organized natural detrementation works (e.g. a certain amount of emissions); if this rate is exceeded, long-term damages of the ecosystem may result. The scarcity of accumulation analyzes a resource or a carrier which is exhausted after a finite number of uses (e.g. fossils, or a landfill).

Social consequences, time pressure, and decreased job enrichment due to monotonous work processes should also be evaluated. Even business knows the wisdom "More haste, less speed." The time span needed to get decision tools into use on a standardized level is much longer than assumed. It took 30 years for the net present value conception to be adopted by the majority of the companies.[14] This process of incubation is necessary, especially for complex facts.

The acceleration trap as an expression of economic consequences has been partially perceived by companies. However, ecological and social consequences are not yet fully recognized.

4 Sustainable Management Instead of Acceleration: Deceleration as a Win-Win Strategy of Companies

In this Section we will show which strategies may be applied to realize deceleration in companies. Deceleration processes will only be accepted if they are win-win strategies, that means, if they have a positive impact on ecological targets and foster company interests at the same time. This is the crucial point, as companies often do not know all their interests, especially if long-term interests are taken into consideration.

First of all, we want to define "deceleration in production": Deceleration in production is the intended retardation of processes on all levels of the value

---

12 Backhaus K. und Bonus H. (eds.)(1997): Die Beschleunigungsfalle oder der Triumph der Schildkröte. 2., erweiterte Aufl. Stuttgart.

13 cf. von Braun C.-F. (1991b): Die Beschleunigungsfalle in der Praxis. In: Zeitschrift für Planung, 2. Jg., Heft 3: 267ff.

14 Weber J. (2002): Betriebswirtschaftliche Instrumente – Segen oder Fluch? In: Kostenrechnungspraxis, 46. Jg., Heft 6: 339–340.

chain which leads to slower material, energy, and information flows. Von Braun uses the image of a water tube for the relationship of process and speed, i.e. its direction, its speed, and its volume.[15] This picture helps to explain the three determinants of the deceleration of processes:

Direction: In which direction does the material flow go, i.e. are resources used or generated?

Speed: How often is there a material flow per unit of time, i.e. how fast are the resources used or generated?

Volume: How big is the material flow, i.e. how many resources are used or generated per process?

Deceleration can be implemented by the consumers, or the company itself. Consumption can be changed by conservatism, leapfrogging, or time investments:

Conservatism is characterized by preferences for goods which can be used for a longer period of time. It is a consequence of experienced negative effects of progress and acceleration. For example, the porcelain company in Meissen nearby Dresden follows a strategy to preserve tried and tested forms and holds a stock of forms dating back to the 18th century.

Shorter innovation and product life cycles combined with price decreases, such as in information technology, may result in slapping one or more technology steps (*leapfrogging*). The consumers decide against the new technology available on the market and focus on future developments (for example slapping one release of a software product). This behavior is influenced by the degree of diffusion and maturity of the new technology and by consumer expectations about upcoming technologies. Leapfrogging is restricted by the fact that capacity and efficiency of the existing technology influence the new technology. Leapfrogging is an alternative if the time span for the adaptation of the system (training etc.) is greater than the time span for the introduction of the new technology.

A third strategy for deceleration by consumers is *time investment.* Time investments mean to abstain from possible time savings. Deceleration is the difference between the time expenses for a time saving alternative (e.g. fast food) and a time consuming alternative (e.g. candle light dinner). *Sufficiency* is a prerequisite for this strategy and turns upside down the so far accepted logic "The faster the better". Other examples can be found in tourism.

Companies can apply two strategies: deceleration trusts and eclecticism:

Deceleration trusts aim at a common deceleration of all competitors of a market. Longer life cycles or innovation cycles are agreed upon. This self-restriction, e.g. in Japanese chip production, is a reaction to threatening efficiency losses and long amortization periods for newly developed products.

[15] cf. von Braun C.-F. (1991a): Die Beschleunigungsfalle. In: Zeitschrift für Planung, 2. Jg., Heft 1: 51–70.

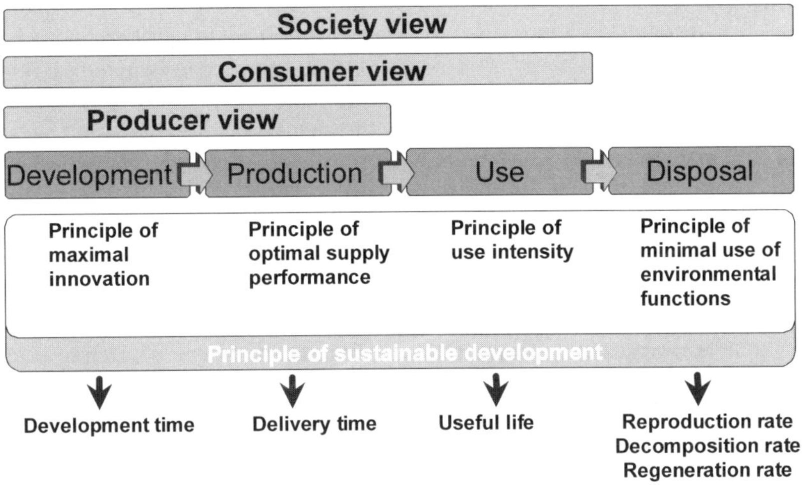

Fig. 2. Principles of time target optimization

Eclecticism – often with a negative connotation – stands for the development of new products out of old ideas. Combined with deceleration, eclecticism stands for the creation of new products and services out of existing components, that are refined, improved and adapted to individual needs. This enables the so far "not fully used" characteristics of existing products and services to be used, and totally new developments become obsolete. This can be combined with conservatism and ends up in an increase in flexibility. Differentiation is the strategy applied here.

Concluding, time target optimization can be structured as follows: The period of development must follow the target of a maximal innovation ability. For the production, the principle of optimal supply performance can be applied. To meet the functions optimally for the use phase, a maximal use intensity must be reached. Last but not least, disposal has to take into consideration with regard to the function of the environment.

5 Is There a Preference for Deceleration? Measuring the Willingness to Pay for Deceleration[16]

In the previous four sections of this paper different theoretical arguments and empirical material on the issue of deceleration, mainly from the producers'

[16] The authors gratefully acknowledge financial support by the "Förderverein der Fakultät Wirtschaftswissenschaften der Technischen Universität Dresden" and thank Yvonne Gerschwitz for support in the realization of the experiments and the preparation of the diagrams.

sphere, have been presented. The question of whether there is also a general preference for deceleration in the population, and if so, how it can be measured exactly, is yet to be answered. There are several approaches which can be used to analyze this question, for instance, demoscopic studies by questionnaires, or econometric studies using statistical data. The procedure used in this study is to measure the preference for deceleration by the agents' willingness to pay for deceleration in laboratory experimental settings.

We designed three experimental settings which we have conducted as class room experiments with students from an advanced course on environmental management at the Technical University of Dresden during the winter term 2003/2004. The **first design** "**Mental Exercises**" tests the willingness to pay for deceleration in a competitive environment where participants could win money by successfully solving a series of mental exercises under time pressure. The individual pay-off of each participant depended on both his or her individual score rank and speed rank. After each one of the six mental exercises every participant could individually decide to continue immediately, or take a break with free refreshments, snacks, and soft drinks offered by the experimentator team. The **second and third** experiment "**Life Cycles of Personal Computers**" and "**More Stress for Higher Income**" were designed as questionnaires. The participants had to imagine a virtual decision situation which was characterized by a trade-off between deceleration, on the one hand, and income, or technological progress and comfort, on the other hand. Of course, we did not communicate the names of our experiments to the participants before or during the experiments.

We will proceed now in the following way. For each one of the mentioned three experimental settings the respective experimental design is first described in greater details (subsection 1), then the empirical findings of the experimental runs are reported. We will present the data as well as quantitative evaluations of the data (subsection 2), and finally we will comment on the experimental evidence (subsection 3). In a résumé we will finally summarize the conclusions from our experiments.

5.1 Experiment 1 "Mental Exercises"

5.1.1 Design

The participants got the following **Instructions**:

> "We will now give you a sequence of six mental exercises – one after the other – each of which yields a certain number of scores which are written on the sheet. After each exercise you can choose to continue immediately with the next one, or to take a refreshment break during which we will offer you coffee, tea, cold soft drinks, and snacks for free. After finishing your exercises we will offer you no more refreshments. Your final pay-off will depend on both the scores you will receive and your speed rank as follows:

Score rank pay-off:
1–3: €4; 4–6: €3; 7–9: €2; 10–12: €1.

Speed rank pay-off:
1–3: €2; 4–6: €1,50; 7–9: €1; 10–12: €0,50.

Your total pay-off will be calculated as the sum of the pay-offs from your score rank and your speed rank. Thus, your maximum possible individual total pay-off is €6, the minimal is €0."

5.1.2 Empirical Findings and Results

The experiment was conducted in March 2004 with 21 students from an advanced course on environmental management at the Technical University of Dresden. A pilot experiment with 23 students of an advanced course on experimental economics at the Technical University of Dresden had been conducted in December 2003 with a slightly different design (cartoons instead of refreshments during breaks, higher possible maximum pay-offs, different pay-off tables) and had shown qualitatively similar evidence (cf. Table 4 and Fig. 6 below). We took care not to mention the issue of deceleration during the course work in the weeks before our experiments.

In the following analysis we will confine ourselves to the March 2004 experiment. On the basis of our empirical findings we are going to analyze the following question which is central to our approach: Is there a willingness to pay for deceleration in the subject pool observable?

The following three tables give a complete account of the empirical observations in this experimental design.

To analyze our central question of whether there is a willingness to pay for deceleration derived from the experiment, we have to first interpret this question in the context of the observable data. Since the number of breaks taken by a subject naturally influences his or her speed rank more or less negatively, we interpret the number of breaks taken by a subject as revealing the subject's individual preference for deceleration. To be more precise, we interpret taking one more break as exhibiting a certain willingness to pay for a worse speed rank and consequently a smaller total pay-off. Thus, the central question of our analysis reads as: How do a subject's breaks correlate with his, or her, total pay-off?

Let us proceed step by step. In a first step we will study how the speed rank correlates with the number of breaks. Fig. 3 below gives a linear regression estimate for this question. The correlation coefficient r is 0.2915, the standard deviation $S_x = 6.06$ and $S_y = 1.05$ (x speed rank, y number of breaks). Fig. 4 shows a linear regression between the speed rank as the independent variable and the total pay-off as the dependent variable. The correlation coefficient is -0.44, the standard deviation $S_x = 6.06$ and $S_y = 1.65$ (x speed rank, y total pay-off).

For the sake of completeness, Table 3 provides the list of the observed score ranks, breaks, and total pay-offs.

Table 1.

speed rank	score rank	total pay-off	number of breaks
1	5	5	2
2	15	2	1
3	19	2	2
4	16	1,5	1
5	13	1,5	1
6	1	5,5	1
7	4	4	1
8	6	4	1
9	18	1	0
10	14	0,5	1
11	2	4,5	1
12	11	1,5	2
13	12	1	2
14	7	2	2
15	9	2	2
16	8	2	1
17	10	1	1
18	21	0	5
19	17	0	3
20	3	4	0
21	20	0	2

Table 2.

score rank	speed rank	total pay-off	number of breaks
1	6	5,5	1
2	11	4,5	1
3	20	4	0
4	7	4	1
5	1	5	2
6	8	4	1
7	14	2	2
8	16	2	1
9	15	2	2
10	17	1	1
11	12	1,5	2
12	13	1	2
13	5	1,5	1
14	10	0,5	1
15	2	2	1
16	4	1,5	1
17	19	0	3
18	9	1	0
19	3	2	2
20	21	0	2
21	18	0	5

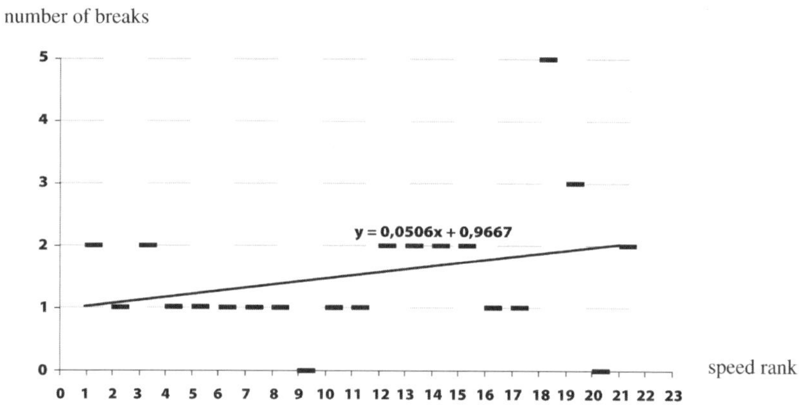

Fig. 3. Experimental Evidence and Linear Regression of Speed Rank and Number of Breaks Taken in the Experiment "Mental Exercises"

Table 3.

total pay-off	speed rank & score rank	total pay-off
1	6 & 1	5,5
2	1 & 5	5
3	11 & 2	4,5
4a	7 & 4	4
4b	8 & 6	4
4c	20 & 3	4
5a	2 & 15	2
5b	14 & 7	2
5c	16 & 8	2
5d	3 & 19	2
5e	16 & 9	2
6a	4 & 16	1,5
6b	5 & 13	1,5
6c	12 & 11	1,5
7a	9 & 18	1
7b	13 & 12	1
7c	17 & 10	1
8	10 & 14	0,5
9a	19 & 17	0
9b	21 & 20	0
9c	18 & 21	0

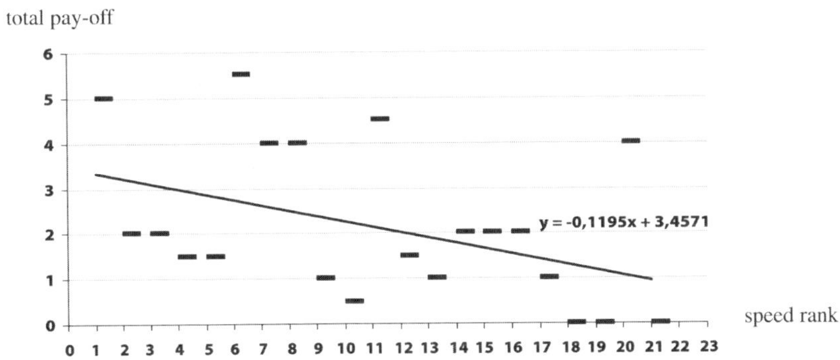

Fig. 4. Experimental Evidence and Linear Regression of Speed Rank and Total Pay-Off in the Experiment "Mental Exercises"

Fig. 5 shows the correlation between the score rank and the number of breaks. The correlation coefficient is 0.44, the standard deviation $S_x = 6.06$ and $S_y = 1.05$ (x score rank, y number of breaks). In Fig. 6, the unbroken

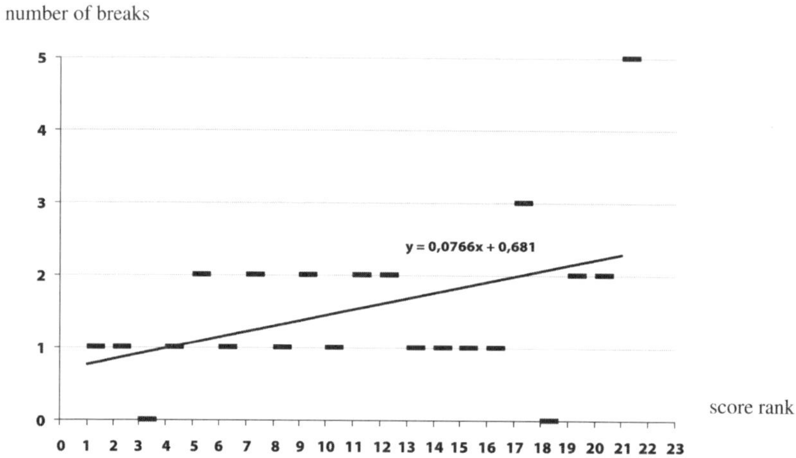

Fig. 5. Experimental Evidence and Linear Regression of Score Rank and Number of Breaks Taken in the Experiment "Mental Exercises"

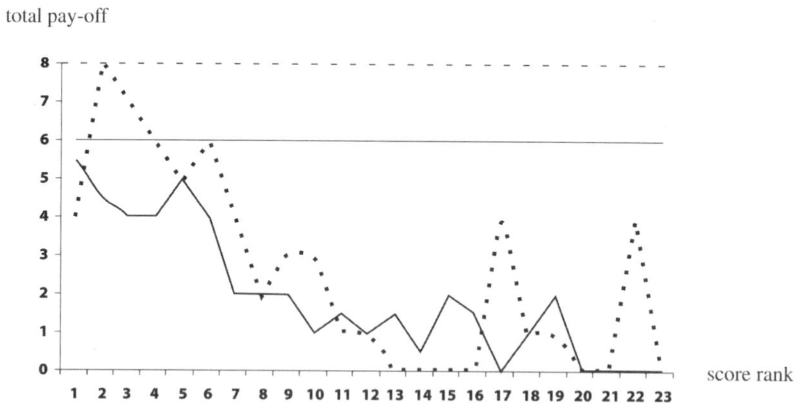

Fig. 6. Experimental Evidence on Score Rank and Total Pay-Off in the Experiment "Mental Exercises"

line maps the data from the March 2004 experiment, the unbroken horizontal

score rank	Total pay-off	
	December 2003	March 2004
1	4	5,5
2	8	4,5
3	7	4
4	6	4
5	5	5
6	6	4
7	4	2
8	2	2
9	3	2
10	3	1
11	1	1,5
12	1	1
13	0	1,5
14	0	0,5
15	0	2
16	0	1,5
17	4	0
18	1	1
19	1	2
20	0	0
21	0	0
22	4	–
23	0	–

Table 4.

line indicates the maximum limit of the total pay-off of €6. The dotted lines indicate the corresponding data for the December 2003 experiment with a modified design, as mentioned above.

5.1.3 Comments

Experiment 1 was an interactive group experiment where the outcome of a participant's decision was dependent on the decisions of the other participants. Let us now look more closely at the central question of how conclusions can be drawn from this design and its empirical evidence about the subjects' possible willingness to pay for deceleration. At first sight, the answer seems to be clear: From the pay-off rule in the instructions it follows that a lower speed rank yields a lower pay-off. Furthermore, a subject's speed rank is naturally negatively influenced by the number of breaks he, or she, takes. Thus, one might conclude from this, that the more breaks a subject takes the larger is his, or her, willingness to pay for deceleration.

Looking at the regression diagram of Fig. 3, the idea that a subject's speed rank is negatively influenced by the number of breaks taken is, in fact, (weakly with $r = 0.291$) supported. The problem with the argument of the previous paragraph is, however, that it is not clear from the outset that a worse speed rank caused by a larger number of breaks actually is positively correlated with a lower total pay-off over the whole empirical data set. This derives from the fact that a subject's total pay-off is composed of two components – the speed rank pay-off and the score rank pay-off. There might be some other effects interfering with the negative pay-off effect of a larger number of breaks, so that in the data there is no positive correlation between a larger number of breaks and a measurably smaller pay-off. Moreover, it is not even clear that a worse speed rank is in fact correlated with a larger number of breaks. On the contrary, it might be the case that the undeniably negative influence of a larger number of breaks is overcompensated by an increased speed of the subject in the succeeding exercise rounds, or by a better quality of solving the mental exercises.

This means, we have to investigate whether there is a positive correlation between a larger number of breaks and a smaller pay-off. For our later conclusions, however, the following statement is important: It appears to be plausible to assume that subjects *expect* that a larger number of breaks causes a smaller total pay-off. Consequently, a larger number of breaks taken by a subject exhibits his or her *self-perceived* willingness to pay for deceleration.

Fig. 4 shows that this, in fact, has been a meaningful assumption: From correlation analysis it follows that a worse speed rank is also, in the whole data set, positively correlated with a lower total pay-off (correlation coefficient $r = -0.44$). Thus, we can conclude that, in the subject pool, there are participants with a preference for deceleration for which they are willing to forego a better performance in speed, and thus to forego parts of their possible pay-off.

Yet, how can we measure the willingness to pay for deceleration? A simple idea is to count the numbers of individually taken breaks. Then, we get the following result: From the maximum possible $5 * 21 = 105$ breaks, the participants, in total, realized 32, i.e. approximately 30%. Only two of the 21 participants took no break at all. 10 subjects took one, 7 two, 1 subject took three, and 1 subject took five breaks during the whole session. From the snacks offered, the sweets were favoured by the subjects, hot drinks, such as coffee, or tea, were less consumed, probably because it took a longer time to drink them than to eat a snack. 17 subjects commented positively on the breaks, 5 subjects had fun with the experiment, 12 wrote that they "felt well", but all subjects emphasized in their comments that the mental exercises meant stress for them.

This evidence is reinforced by the fact that there is not only a positive correlation between a worse speed rank and a larger number of breaks, but also a positive correlation between a worse score rank and a larger number of breaks a subject took, as Fig. 5 shows. An explanation for this could be a negative effect of breaks on a subject's concentration and ambitious attitude towards the whole experiment. The other direction of causality, however, may also be true, which means that there is a self-preselection effect of subjects with low ambition which is coupled with a greater inclination to take a break.

5.2 Experiment 2 "Life Cycles of Personal Computers"

5.2.1 Design

The participants were given the following **Instructions**:

> "Imagine you need a PC/laptop of a middle technological quality for professional reasons and you have to pay for it with your private money. Which one of the following two technological development scenarios A and B for PCs/laptops of a middle technological quality in the following diagram would you prefer?
>
> scenario A = full lines
> scenario B = dotted lines

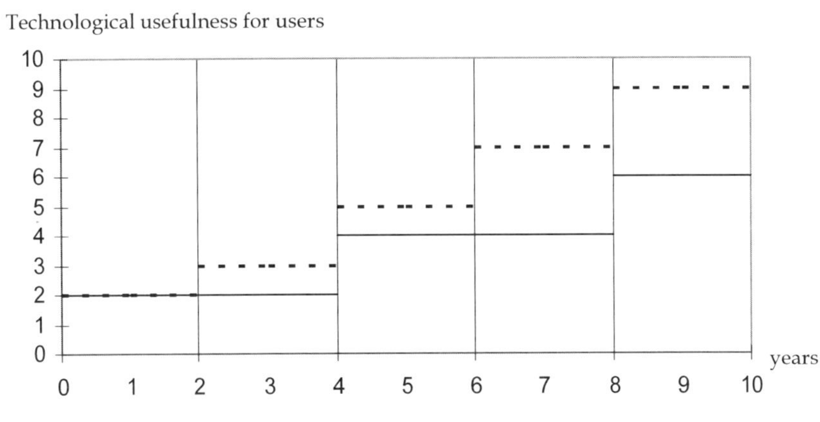

Fig. 7. Instruction Scheme of the Experiment "Life Cycles of Personal Computers"

> Please, describe the reasons for your decision."

5.2.2 Empirical Findings and Results

The experiment was conducted in March 2004 with 21 students from an advanced course on environmental management. Scenario A stood for the decelerated, scenario B for the accelerated case. The empirical findings were as follows:

Table 5.

No. of subject	chosen scenario	
1.		B
2.		B
3.		B
4.		B
5.	A	
6.		B
7.		B
8.	A	
9.	A	
10.	A	
11.	A	
12.	A	
13.	A	
14.	A	
15.	A	
16.	A	
17.		B
18.		B
19.	A	
20.	A	
21.	A	

The distribution of absolute numbers of choices looks like Fig. 8a, the distribution of relative numbers (percentages) of choices is shown by 8b.

5.2.3 Comments

Experiments 2 and 3 were not interactive group experiments like experiment 1, but questionnaires. In experiment 2, our findings show an even stronger preference for deceleration than those of experiment 1: almost two thirds (61.9%) of the subjects chose the decelerated A-scenario.

From the answers to the last question, asking why the participants chose scenario A, or B, respectively, we have learnt the following: Most participants had understood the decision situation properly and commented on their individual decision in a comprehensible way as expected: A-type subjects preferred fewer changes of their laptop over the course of time and were not interested in accelerated technological progress since, in their opinion, many functions of a computer are not used by average users. B-type subjects, on the other hand,

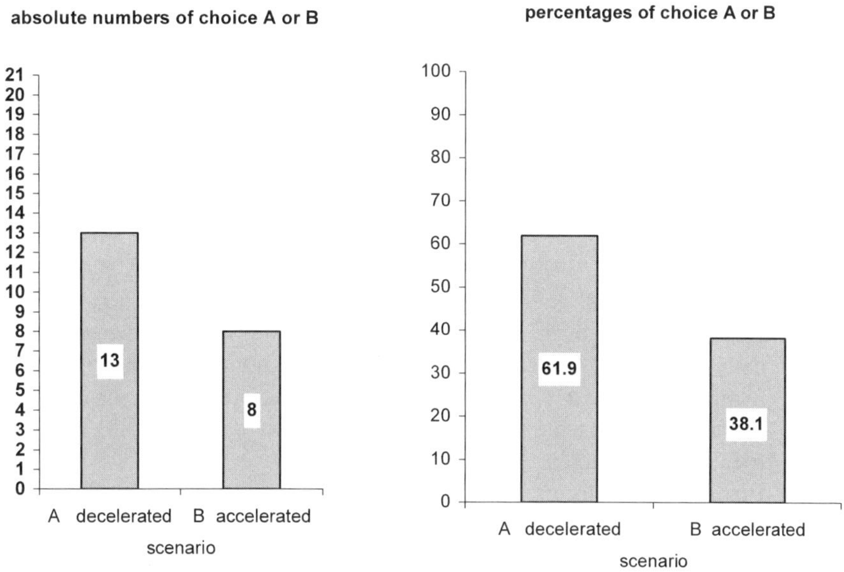

(a) Absolute Numbers (b) Percentages

Fig. 8. Experimental Evidence on the Preference for Deceleration in the Experiment "Life Cycles of Personal Computers"

stressed the necessity of a high technological standard of a laptop deployed for professional use. In both, the A-choice- and the B-choice-party, there were also some subjects who did not comply with the instructions, but referred to considerations which were not included. Typically, subjects of this type choosing the accelerated scenario B argued that they would prefer to lease the laptop/PC instead of buying it, as the instructions say. Subjects of this erroneous type who chose the decelerated scenario A typically argued that they would use a laptop for private purposes only, though in the instructions we clearly told them that the laptop was needed for professional reasons.

In the design of Experiment 2 we deliberately did not speak about prices for PCs, or laptops. In a former pilot experiment of this type we found that, if we did speak of prices, the participants would primarily calculate their monetary advantage from the slower or faster development scenario. The aspect of deceleration became secondary in their decision. This might be interpreted as a low significance of the deceleration issue from the subjects' point of view. Following another interpretation, which in our eyes is more relevant, one could argue that students of business administration are specially trained in calculating monetary advantages. Thus, they would perceive our decision situation as one of optimizing the monetary pay-off instead of taking the "soft" criterion of deceleration into account.

5.3 Experiment 3 "More Stress for Higher Income"

5.3.1 Design

The participants got the following **Instructions**:

> "Imagine you have successfully passed the final exam in business administration at the Technical University of Dresden and you have already applied for a professional position in several firms. Two firms, A and B, will accept you:
>
> 1. Firm A primarily expects you to be flexible and not geographically restricted, to accept irregular working hours, including being available to work also on Sundays and holidays, if necessary, to be flexible with your holidays and always to accommodate to the firm's requirements.
> 2. Firm B primarily expects you to be flexible and open-minded for further qualification and offers you regular working hours. You can furthermore plan your holidays in coordination with your colleagues in advance.
>
> Which one of the two firms, A and B, will you choose in the each one of the following three cases?:
>
> (1) You will earn €70.000 per year in firm A, and €40.000 in firm B.
> (2) You will earn €60.000 per year in firm A, and €40.000 in firm B.
> (3) You will earn €50.000 per year in firm A, and €40.000 in firm B.
>
> Please, write down the reasons for your decision."

5.3.2 Empirical Findings and Results

The experiment was conducted in March 2004 with 24 students from an advanced course on environmental management. Firm A stood for the accelerated, firm B for the decelerated case. The participants answered in the following way: There were four different patterns of answers observable in our experiment:

(1) AAA (2) AAB (3) ABB (4) BBB

As one should expect, the empirically observed answer patterns are "monotonic" with respect to the intruding of "B" from the right end of the triple. Why did patterns like BAB, or BBA for instance, not occur in the empirical findings? The answer is clear: Due to the given sequence of the three cases (1) – (2) – (3) in the experimental design, any answer exhibiting a non-monotonic pattern, like BAB, or BBA, would be inconsistent and irrational since the incentive to choose the lower income of firm B is the greater the smaller the income difference is, i.e. the higher the case number.

Table 6.

Running number of participant	Case 1	Case 2	Case 3
1.	B	B	B
2.	B	B	B
3.	B	B	B
4.	B	B	B
5.	A	A	A
6.	A	A	B
7.	A	A	A
8.	A	B	B
9.	A	A	B
10.	A	A	B
11.	B	B	B
12.	A	B	B
13.	B	B	B
14.	A	A	B
15.	B	B	B
16.	A	A	B
17.	B	B	B
18.	B	B	B
19.	A	A	B
20.	B	B	B
21.	A	B	B
22.	B	B	B
23.	A	A	B
24.	A	A	B

The percentage of each one of the four observed patterns is presented in Fig. 10.

The percentages of firm A choices or firm B choices in each of the three cases (1) to (3) are presented in the following Fig. 11.

5.3.3 Comments

In this experiment, we also find a clear preference for deceleration. Almost half of the subject pool (46 %) chose the decelerated working conditions of firm B in all three relative income scenarios. In two of three income scenarios, the majority chose alternative B – foregoing a significantly higher amount of income (€10,000 or €20,000 p.a.). Even in the first case, where the distance between income in the accelerated and the decelerated scenario is €30,000, the number of A- and B-choices is almost equal (54.2% chose the accelerated scenario A) whereas in the case with the smallest income difference of €10,000 p.a. almost all subjects chose the decelerated scenario (91.6%). This means

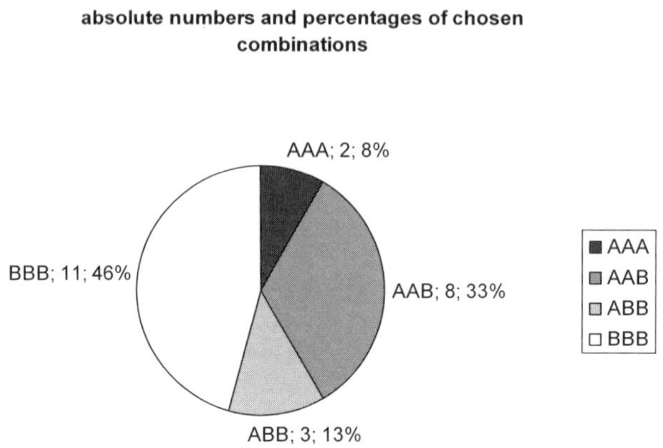

Fig. 10. Experimental Evidence on the Preference for Deceleration in the Experiment "More Stress for Higher Income" (Cake Diagram)

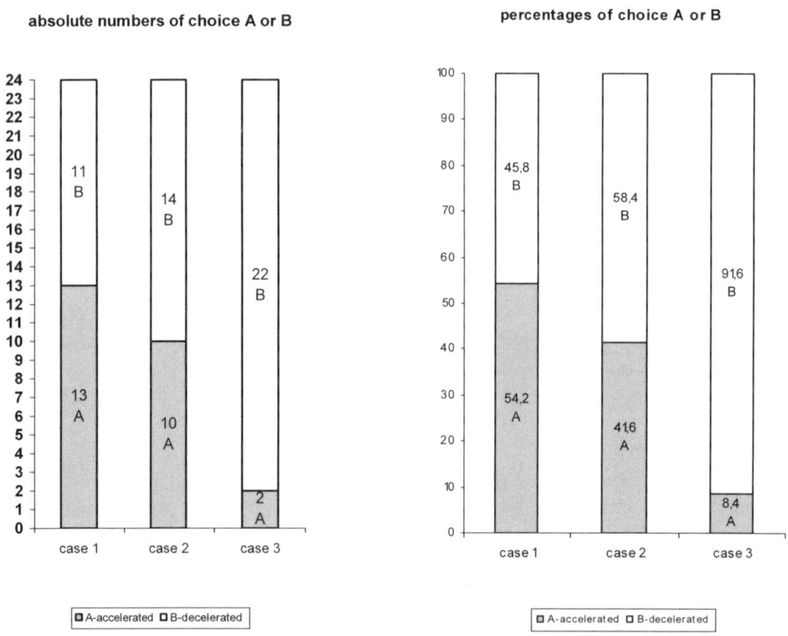

(a) Absolute Numbers (b) Percentages

Fig. 11. Experimental Evidence on the Preference for Deceleration in the Experiment "More Stress for Higher Income"

a notably high willingness to pay for deceleration which, moreover, increases with opportunity costs getting lower.

The comments by the participants illustrate and corroborate this revealed monotonically increasing willingness to forego the higher income alternative A for the alternative B. As expected, the main reasons for choosing the decelerated alternative B were more leisure time and more time for family and social activities, less working stress, and better chances for further education.

5.4 Résumé of the Experimental Evidence

We have designed and conducted three laboratory experiments for a better understanding of whether subjects have a preference for deceleration at all, and if so, how the preference for deceleration can be measured. We tried to analyze these questions by confronting subjects with different trade-off situations between an accelerated and a decelerated alternative and different opportunity costs of the decelerated alternative. Only the first one of our three experimental settings was an interactive group experiment where the personal outcome of each participant was interdependent of the decisions of all other subjects. Experiments 2 and 3 were conducted using questionnaires.

However, as we have seen in our findings, in all three of our experimental settings, we observed a clear preference of the subjects for deceleration. As could be expected, the subjects, throughout all of our experiments, showed a preference for deceleration with an increasing willingness to pay with decreasing opportunity costs. These opportunity costs were: a possible higher speed and score rank and, accordingly, a higher monetary pay-off in the first experiment, a faster increasing technological usefulness of PCs/Laptops in the second, and a higher income in the third experiment. Deceleration was represented by refreshment breaks in the first, a slower increase of technical usefulness for users of a new technology application in the second, and more comfortable time management and conditions on the job in the third experiment.

One usual criticism of laboratory experiments in social sciences pertains to the choice of the subject pool. In our experiments, the subject pools indeed were formed somewhat selectively by students from an advanced course on environmental management at the Technical University of Dresden. The criticism, consequently, might be that young people without job and family responsibilities will, of course, have a greater willingness to pay for a more decelerated way of living than people with a job and raising kids, for instance. Or in other words, students normally experience a phase of their lives in which the social obligations are particularly low compared with later life phases, and thus may tend to underestimate income and to overestimate their own well-being.

We are certainly well aware of the fact that the selection of our subject pool might have had a biasing effect in the direction of greater willingness to pay for deceleration than subject pools from other parts of the population.

But we are convinced that, in any case, it is interesting to see what young people, who are passing academic studies and consequently have a great chance of later belonging to the elite of the society, think about the question of deceleration. Nevertheless, it is desirable to repeat the experiments with subject pools selected from other parts of the population, for instance parents, workers, employees, independent businessmen and -women, and also high school pupils. The latter group is of particular interest since they, like university students, will carry over their present preferences with respect to acceleration/deceleration, in some way or other, to the future and will accordingly shape the future societal and working reality.

6 Summary and Outlook on Future Research

The central aim of our present study has been to verify that deceleration is not only a fashionable issue of current public discussion, but also a real measurable phenomenon. In our study, we show two results by conceptual considerations and empirical findings: Deceleration is a win-win strategy for sustainable management, and furthermore, there is significant experimental evidence of a preference for deceleration in the society. Companies will have to face the challenge to merge time targets of consumers and the environment with their own targets in order to reach time target optimization. Besides theoretical analyses concerning the implementation and the effects of deceleration, empirical studies will become more relevant. Together with case studies, experiments which allow the analysis of effects in laboratories – so to say "under a magnifying glass" – will become more important. Due to the rapid development of experimental economics and existing strategic games, science is well prepared for this new task. For this, the vital field of experimental economics and the older business planning games provide a research infrastructure, which, however, to our knowledge has not previously been used for analyzing the issue of deceleration.

The experimental evidence we have found here must, however, be corroborated by later repetitions of our experiments using different subject pools and probably also new treatment variants. We can, however, already conclude from our experiments here that there is a significant preference for deceleration in the society which manifests in quite different contexts, i.e. experimental settings, and which can furthermore be measured by the agents' willingness to pay in trade-off-situations where more deceleration has certain opportunity costs. By this we mean values which are generated by acceleration: more income, faster technological development, less production time, more output, and so on. It has been the main concern of this study to analyze whether a continual increasing and intensifying of these traditional targets truly generate increased utility and wealth, which they are assumed to do. In fact, our findings strongly support the argument from the discussion on the topic of happiness that traditional economic targets, like those just mentioned, must be reinterpreted more comprehensively to maintain their function as meaningful notions of human life.

This article shall end with a legend about Pablo Picasso. Being asked by a collector to paint a picture for him, Picasso drew some strokes within one hour and said: The price is $100,000. The collector thought this to be impertinent and complained: "For one hour of work that much money?" But Picasso replied: "That didn't take me one hour, but 80 years." Of course, he was correct. He collected 80 years of experience and created a brand which maintains its high value even now.

Comment: Deceleration as a New Paradigm of Economic Science?[1]

Fritz Reheis[2]

[1] Translated from German by Dipl.-Vw. Christoph Heinzel, Technical University of Dresden

[2] Branigleite 19, D-96472 Rödental b. Coburg.
fritz.reheis@t-online.de

The concern of Günther and Lehmann-Waffenschmidt's contribution is remarkable: the opening up of the discourse of business management and economics for the subject of "time, acceleration, deceleration". Moreover, the authors conduct their analysis not only in the context of time-management strategies, but also against the background of the meanwhile widely accepted guiding principle of sustainable development. Hence, the study is, beside the increase of efficiency on the level of the company, about the question of the global conditions of life and survival. Thus, the authors tackle a question which is relevant in terms of economic and social ethics because it introduces, beside the criterion of efficiency, the criterion of justice into the business management discourse. Therefore, the authors address a holistic concept of quality of life and happiness. In a first step, I will examine whether Günther and Lehmann-Waffenschmidt are consistent with their own claims. In a second step, I will outline my own considerations towards a productive continuation of the discourse.

1 Consistent with Their Own Claims?

The general question whether deceleration may become "a new paradigm of economic science"[3] is divided by the authors in the beginning of the text into four single questions which structure the contribution.

The question on the *causes*[4] of acceleration is answered by presenting a multitude of well observable real-world facts which are presented as factors

[3] Translator's note: Günther and Lehmann-Waffenschmidt initially ask whether there is a preference for deceleration in the society and whether deceleration can become a paradigm in *business management.*

[4] Translator's note: Günther and Lehmann-Waffenschmidt speak of "reasons".

of acceleration and located on three levels. The authors present a detailed and broad bundle of factors for acceleration in economy and society comprising the sides of both, producers and consumers. It has to be questioned, however, whether thereby "causes" in a narrower, i.e. philosophical, sense are presented. According to the dominant economic discourse, the explaining factors are usually understood to be independent variables within a larger economic model from which the explanandum is deduced as dependent variable. For a complete explanation of all those conditions, which cause the acceleration of the economy and of life, it would therefore, within the discourse of time and sustainability, be necessary to throw light not only on the presence of the single factors, but also on the suitability of the whole background model of the economy and thus on the relationship and origin of these independent variables. For instance, there are fundamental interrelationships between the exponential compound interest effect of the growth of monetary wealth, which Günther and Lehmann-Waffenschmidt state on the macroeconomic level, and the companies' endeavour to diminish costs by "time-related strategies" via shortened capital use times on the microeconomic level. Thus, the classification of the factors as "independent variables" of the real world would be patently inept. More concretely: If interests were equal to zero, or even negative (so-called "rusting" money), neither the first, nor the second factor would exist. As such, the routinized working of the dominant economic science with dependent and independent variables, where the latter are usually subsumed under the "ceteris paribus" clause, also prevents Günther and Lehmann-Waffenschmidt from further questions at this point. Such questions lead to the dispute on Silvio Gesell's concept of free money ("Freigeld") which, in itself, cannot be discussed without fundamentally questioning general equilibrium theory and its methodological individualism [4], and, hence, they have to be part of a (descriptive-analytical and prescriptive) context of discourse in social and economic philosophy.[5]

Also concerning the *consequences* of acceleration, the contribution contains apt observations on the micro and macro levels. The ecological consequences on the side of the sources and sinks of nature's household and the business consequences on the producers' and consumers' sides are presented in an impressing manner. In my opinion this presentation is, however, not complete. First, the second Section lacks the motivational level which was contained in the first Section of the paper. Presumably, this is related to the fact that questions of motivation are not considered to need further clarification in the basic model of general equilibrium theory with its exogenously given preferences. Although at no place in the text the authors mention such a

[5] Editor's note: The question of „Freigeld" in a socio-economic context has been tackled in two articles by M. Lehmann-Waffenschmidt: Vision und Kritik der modernen Wirtschaft in Goethes 'Faust'. In: Mahl B. und Loerke T. (Hrsg.) (2005): Faust-Jahrbücher, Band I, Francke Verlag, and in: Geld, Entgrenzung und Glück. Das Psychogramm unserer Zeit in Goethes 'Faust'. In: Karmann, A. und Klose, J. (Hrsg.)(2006): Geld regiert die Welt.

general equilibrium model, the basic notions of utility and scarcity, the rationality principle, demand and supply, price and market coordination, cannot be used in a meaningful way without such an "implicit" background model. If one asked for the independent variables for dependent preferences, one would quickly touch upon the motivational reasons from religious and depth psychology, mentioned unfortunately only in the Section on causes, but also purely economic ones for cooperative behaviour, like e.g. the double formation at the working place by formal and material renunciation of sovereignty of action. Concretely: A worker, who constructs cars on foreign command wants to buy a car on his own decision as a consumer – and supposedly even gets into debt for it. It can be presumed that rising heteronomy and pressure at work result in rising need for compensation. Secondly, beside ecological and business consequences the acceleration of economy and society is considered to be also related to socio-economic ones. There are convincing hints for a deepening gap between the fast and highly productive part of the population on the one hand and the slow and less productive one on the other in the development of capitalism. This macroeconomic distributive effect of acceleration, is completely absent in Günther and Lehmann-Waffenschmidt's contribution.

The third of the authors' introductory questions, regarding the *contribution of deceleration to a sustainable economy*, and the forth one on the *implementation of the idea of deceleration*[6] are treated together although they refer to different problems. Concerning the question of the contribution of deceleration to sustainability three things come into one's mind. First, one would like to know how the three accepted dimensions of sustainability (ecological, economic, social/cultural) are integrated in the notion of sustainability used by the authors. For, in the scientific community, there is only a consensus on a basic terminological-normative level, but not for more concrete cases. That means, "harder" and "softer" notions of sustainability are distinguished, up to 60 definitions of sustainability have been counted (cf. [1, p. 16–29]). In which way, however, deceleration shall be shown to contribute to sustainability if there is not such a definition? By the way, this definition would have to be considerably fundamental so that the level, or terminology, of economics can be subsumed in a systematic, or logical, manner[7]. Second, the term "deceleration" should also be defined with respect to reproduction, communication, etc. The illustration by means of the water tube further ne-

[6] Translator's note: Günther's and Lehmann-Waffenschmidt's fourth question reads as "Is there a preference for deceleration in society, and how can it be measured?" However, in Section 4, they also consider strategies to realize deceleration in companies.

[7] Translator's note: In Sections 3 and 4 Günther and Lehmann-Waffenschmidt present and discuss the consequences of acceleration at length (for the notion of 'sustainable' see particularly Section 3.1). The term "sustainable management" is used in Section 4 in the usual sense of environmental business administration. Particularly, as the authors point out, acceleration, or growth, processes cannot be sustainable in a finite world by the literal sense of the notion 'sustainable'.

glects the fact that many massive throughputs of matter and energy are well sustainable, namely if they constitute circular processes. Only if production and reproduction/consumption are terminologically separated, one can speak of deceleration of production. In fact, however, as a glimpse in any biology textbook teaches, most constructive processes are destructive processes at the same time, thus sources and sinks are usually interlinked in a circular way. Third, from the outset, Günther and Lehmann-Waffenschmidt restrict their question concerning the possible contribution of deceleration to sustainability to the win-win criterion. As such, contributions exogenous to the market, which are e.g. induced by incentive systems or bans and permissions, are per se excluded. Presumably, however, most situations of decision in the context of deceleration and sustainability are situations of prisoners' dilemma which can only be overcome by an extension of the communicative opportunities and by setting new basic conditions.

The three experiments presented by the authors are understood as first steps towards an empirically supported answer to the last of the initial questions which aims at the implementation of deceleration. The experiments with university students from Dresden show that many people in many situations wish for deceleration and are also willing to pay for it. Already in the beginning of the 1990s four Germans out of five complained that everything was changing too fast, they preferred things to be somewhat less hurried [3, p. 38]. As far as the question of practical implementation of the idea of deceleration is concerned, however, three things are crucial. First: Where does the purchasing power come from which only permits the demand for deceleration? General equilibrium theory refers to income from salaried labour or wealth which is, in itself, explained on the basis of preference theory. The questions concerning the structural obligation to work, or not to work, are victims of abstraction, just as the questions concerning the distribution and leaving of wealth. Second: Who will produce the supply of goods and services which meets this demand with purchasing power if it becomes clear that a general cultural change towards the deceleration of production, consumption, communication, etc. threatens anticipated profits and economic growth in general? According to general equilibrium theory in the tradition of Adam Smith via Léon Walras to Paul Samuelson demand creates its own supply. However, it is only in this way that a democratic character can be attached to the market economy, as expressed in Samuelson's famous phrase of the "consumers' dollar votes". In fact, outside the model world, in real life, it is, the other way round in many cases. Third: How do the temporal patterns and strategies of those markets where natural resources, human capital, and goods are traded behave to those markets which are concerned with the buying and selling of capital? Günther and Lehmann-Waffenschmidt correctly speak, at this point, of the compound interest effect as an explosive force.

Generally, it should be analyzed in future research how experiments can, beyond the willingness to pay for deceleration, also show the possibility, or probability, of win-win strategies which are favourable to sustainability, thus referring to equilibrium states which guarantee not only the individuals' optimal utility on the sides of both, supply and demand, but which also meet the sustainability criterion.

2 Continuation of the Discourse

My thesis is: If one is concerned with time, acceleration, deceleration and the relationship between non-sustainable and sustainable forms of economic action, then one must leave the narrow framework of the classical- neoclassical model world with its exogenous preferences and exogenous scarcities and choose a materialist-evolutionary perspective. It is only from this starting point that deceleration can become a paradigm of economic science.

For a long time, the *assumptions of the classical-neoclassical model* have been proving absurd. As a matter of fact, this model claims the market to be the best possible institution of coordination in the light of the pursuit of maximum utility of man and the scarcities of nature. The sophistication of the model has often been advanced, among others following an ecologically motivated criticism. Still, these sophistications of the neoclassical workshop equal ad-hoc hypotheses in the philosophy of science and are to be compared in the history of science to the efforts with which supporters of the Ptolemaic image of the world defended themselves against the Copernican [5, 8]. The more capitalist, industrial societies socialize working processes, and the more complex and, at the same time, faster the development of new products is, the more questionable seems to me the notion of preferences being independent from contemporary as well as past social experiences of the individual market subject that decides on a certain demand behaviour out of her own "free will". Rather, in terms of space, her preferences are a priori inseparably connected to those she perceives in her environment and have, in terms of time, a concrete formation history which is in part a product of market processes themselves. The more the exploitation of capital is linked to the implementation of a certain consumption and wealth model, and the more sophisticatedly the so-called leisure time is marketed in the media society, the greater is the danger that individuals are exploited for foreign purposes. Only at the cradle of the market society it may have been that the farmer on his way to the city already brought his preferences with him in his mental rucksack: exact ideas of the properties of the axe he wanted to buy. It seems, however, that the late-bourgeois individual has to be more and more described as the prototype of a socialized being.

An analogue process of condensing and integration can be found on the side of nature. The farther interventions reach into nature, the longer the spatial and temporal causal chains become, the less accurate become axioms which abstract from space and time in such a strong way. It is more and more difficult to theoretically isolate the single scarce goods, becoming the object of the market process, from their integrative relationships, and thus, temporally, their genesis is itself a result of the market process. It is only under the mentioned, idyllic conditions that the forest, in which our farmer cuts a tree with his new axe, is ecologically really only marginally spoiled by this intervention. We know today that a forest industry by using the forest as wood plant and by ignoring the complex interplay of micro organisms, small animals and small plants, up to the climate system, results in catastrophic effects.

In sum: The spatial individualization/parcelling-out strategy with its basis in the modelling technique and the temporal deprocessualization/dehistorization strategy concerning the input side of the market model must be judged as a dangerous anachronism, or as an ideology, if the real effects of condensing and integration gush out more and more massively from all angles of the market system. The term "external effect" ignores this fact completely.

The dominant mainstream of economic science has led us to a double aporia: Our daily efforts for a better life have led to a giant economic rise (measured in terms of the social product), ecologically accompanied, however, by a dangerous decline (measured in entropy). The self-supporting and accelerating processes with the characteristic exponential shape of the curves (cf. the compound interest effect) are, for the biologist, signs of an existential danger of life (cf. the cell degeneration effect). In order to solve aporias, one needs a higher vantage point, a larger frame of reference. This must contain a priori the economy and ecology as well as subjective and objective elements. In my opinion, these conditions can be met in a combined *materialist and evolutionary approach*. Such an approach would have to depart – in exact opposition to the neoclassical model – from the purely objective facts of space and time, and show the emergence of the purely subjective, reason-guided free will. It is only here where one may think about the possibility of free preferences which are, according to the dominant teaching, simply axiomatically assumed. A materialist-historical approach would virtually have to nestle against the factual logic of reality. To put it differently: The analysis must start with the ecological conditions of economic action instead of conceding them by ever new statements of externalities a posteriori. The lonely *homo oeconomicus* and the not less lonely scarce resource have to be put finally on the rubbish heap of the history of economic thought.

The "ecology of time" [6, 7] offers a basis for such a materialist-evolutionary approach. It addresses the time of nature, of culture, of society and of the individual in a unified whole. From the beginning it aims at the questions of the sustainability of economic action and of life. For the operationalization of sustainability it examines the system's own times of processes, their permanence and cyclicity, and pays particular attention to the limits of elasticity and the error-friendliness of interventions. Against this background, the question on the causes and consequences of acceleration appears in a somewhat different light: it regards the dynamics of money, its inherent programme time, which forces the acceleration upon the world in manifold ways and which leads to an encompassing desynchronization and destruction of own times. It is exactly in this framework that measures of deceleration can be scientifically deduced. It is a different question, how to implement them. There is a wide field of options for deceleration, ranging from obligatory state norms for the treatment of nature – including the provision of the means for this which also need to be politically caused – over novel incentive systems to enhance the respect of own times to purely endogenously occurring win-win strategies in markets. We should study and try them immediately.

References

1. Final report of the Enquête-Kommission "Schutz des Menschen und der Umwelt – Ziele und Rahmenbedingungen einer nachhaltig zukunftsfähigen Entwicklung: Konzept Nachhaltigkeit. Vom Leitbild zur Umsetzung," BT-Drucksache 13/11200, June 26, 1998
2. Held M., Hofmeister S., Kümmerer K. and Schmid B. (2000): Auf dem Weg von der Durchflußökonomie zur nachhaltigen Stoffwirtschaft: Ein Vorschlag zur Weiterentwicklung der grundlegenden Regeln. In: GAIA 4/2000: 257–266
3. Piel E. (1993): Die Deutschen fürchten Stress und Langeweile. In: NATUR 11/93
4. Reheis F. (1986): Konkurrenz und Gleichgewicht als Fundamente von Gesellschaft. Interdisziplinäre Untersuchung zu einem sozialwissenschaftlichen Paradigma, München, Berlin
5. Reheis F. (1995): Ökologische Blindheit. Die Aporie der herrschenden Wirtschaftswissenschaft. In: Das Argument 208 (1995): 79–90
6. Reheis F. (1998): Die Kreativität der Langsamkeit. Neuer Wohlstand durch Entschleunigung. 2. Aufl., Darmstadt
7. Reheis F. (2002): Ökologie der Zeit. Zum angemessenen Umgang mit natürlichen Ressourcen. In: Backhaus J. und Helmedag F. (Hrsg.): Holzwege. Forstpolitische Optionen auf dem Prüfstand, Marburg: 139–155
8. Vogt W. (1973): Zur Kritik der herrschenden Wirtschaftstheorie. In: Vogt W. (Hrsg.): Seminar: Politische Ökonomie. Zur Kritik der herrschenden Natioanlökonomie, Frankfurt/Main: 179–205

Assessment Criteria for a Sustainability Impact Assessment in Europe

Raimund Bleischwitz

Stellv. Leiter Forschungsgruppe "Stoffströme und Ressourcenmanagement", Wuppertal Institut, Postfach 100 480, D-42004 Wuppertal and visiting professor at the College of Europe, Bruges (Belgium).
raimund.bleischwitz@wupperinst.org

1 Introduction

The notion of sustainability is well established. It is increasingly used not only in the environmental field, but also in the realms of social security systems and financial stability. Though this might be seen as elusive, it leads to acknowledge the demand for cross-cutting policy approaches. Such a demand is also characterized by a shift from government to *governance*, meaning that institutions and actors other than the state become more important. It is claimed that not only the acceptance of those actors is pivotal for any implementation of policies, but that corporate and societal actors play a role in policy formulation, precautionary measures, and innovation. General drivers of the change from government to governance can be identified as follows [13, p. 202f]:

- Dissatisfaction with environmental regulation: Concerns about implementation costs in bureaucracy, compliance costs in industry, and the disability to get through to small and medium-sized enterprises as well as to private households were raised in most OECD countries during the 90s. The call for cross-cutting, flexible approaches spurring innovation [8] followed.
- Shift in the regulatory debate: Previous arguments about the merits of state-driven policies were perceived less attractive than their counterparts. A governance 'turn' in most OECD countries spurred privatisation programmes and enabled private activities alike. Important to note, the sustainability paradigm has not been directly aligned to these debates, but has also highlighted some limitations of traditional regulation [4].
- Market integration and influence of the European Union: Most OECD countries and, in particular, the European Union have intensified their efforts for market integration, i.e. an harmonization of legal institutions with the aim of reducing transaction costs for international business operations while coordinating economic policies. These efforts almost automatically

have required a more systematic thinking about appropriate levels of action and about cross-cutting approaches. With regard to sustainable development, the EU's regulatory kitchen is not yet designed by a clear concept, but rather by experiments in different directions of economic incentives, voluntary agreements, eco-labels, and formalized planning tools.

Given this debate, the following article will develop assessment criteria for sustainability policies. The assessment criteria are geared towards what is called "regulatory impact assessment" by the European Commission (COM/2002/0276 final). Another reference can be made to the provision of services which the recent annual report of the World Bank (2004) puts great emphasis on. The proposition is that such impact assessment ought to entail far more than setting a one-off framework for internalising externalities, or an upgrading of environmental indicators. Our research questions are as follows: How can policy fulfil its task – in awareness of its own knowledge deficits, and at minimal regulatory cost? How can policy minimize the costs of market coordination and generate new knowledge for solving problems in a dynamic world? Policy in this regard can be analysed as a collective learning process in which the subsidiary effects concomitant with the shift of government tasks to market-based institutions play a major role (Fig. 1).

Our article will test the feasibility of innovation research, new institutional economics, and evolutionary economics and will bring their theories to scientific analysis of policies and governance systems. These approaches [7, 23, 33] hypothetically prove a helpful device for analytically integrating the necessary theoretical elements in a way not permitted by other approaches. Public choice theory assumes that political units act on shared motivations and have an awareness of interest groups, leaving little room for learning processes. Welfare economics, in turn, runs into analytical problems when it comes to bridging knowledge deficits and second-best options. While New Institutional Economics is certainly helpful in examining political processes, it may take insufficient account of learning processes and dynamic change. The regulatory approach offers interesting points of contact with evolutionary economics: as the borderline between state and economy gets blurred, the stabilizing function previously exercised by the legal framework increasingly shifts to adaptive processes of institutions located somewhere between market and government. Our approach addresses the deficits described above. We expect our scope to be helpful in applying theoretically derived principles of open development and experimentalism operationally. Finally, policy-oriented conclusions will be drawn.

2 Legal Frameworks and Institutional Reforms

Reforming and designing institutions are functions in policy which can hardly be conceived of as one-off framework setting activities. According to Pierson [24] and Wegner [32], the fundamental difficulty of setting a legal frame lies in forecasting side effects and the reactions of innovative market players. Moreover, there is a causal relation between framework-setting and political lobbying: the stronger the regulating administration elaborates a specific regulation, the fiercer the interest groups' efforts to influence the decision in the making. A realistic compromise entails granting exemptions of the kind Posen [26] has compared on an international level.

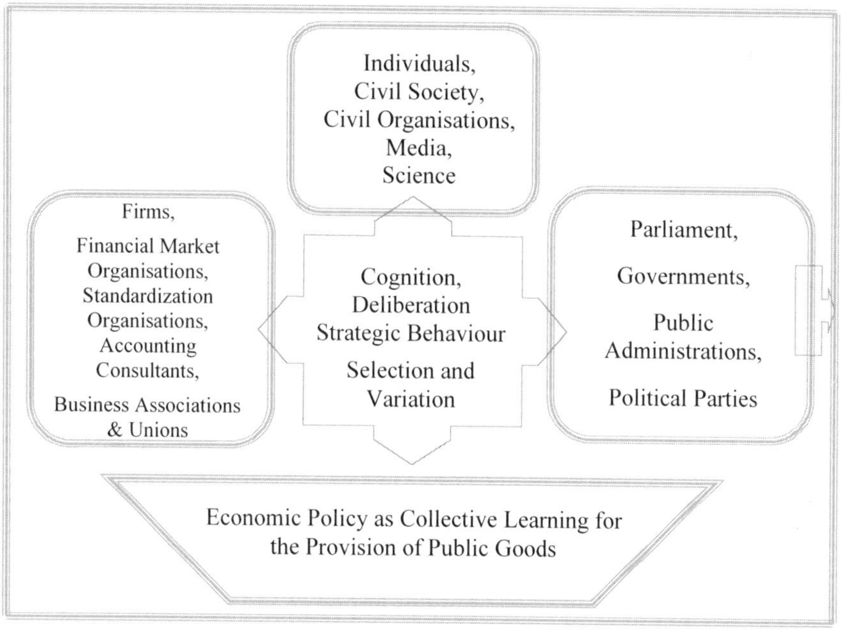

The double lines refer to institutions. Organizations act in an institutional setting. Political decision-makers decide upon outer institutions (e.g. a constitution). Source: own compilation.

Fig. 1. An evolutionary approach to economic policy

This, however, results in high costs for eliminating exemptions. According to Dixit [7, p. 146], the cost of change depends on the rationality of the actors involved. Any well-meant change in regulation can therefore result in higher transaction costs. In other words, a change in the regulatory framework is only beneficial if the costs of change are lower than the costs of retaining existing regulations. Some inconsistencies between old and new regulations will only emerge as the latter are implemented. While according to Witt [33, p. 11], evolutionary economists may even welcome this process because it submits institutional innovations to a kind of stress testing, it also means assessing the effect of a regulation before it passes into law involves prohibitive costs.

From an evolutionary perspective, the rationality of one-off regulations is relative. According to the "knowledge-creating competition" model [14], new framework conditions only have an immediate effect on the business selection environment if they include binding regulations and penalize non-compliance. Government regulation for business does not automatically affect financial institutions and standardization authorities. Indirect stimuli coming from formal institutions – such as the aims and principles underlying laws – require reinforcement through other factors since dynamic and open market processes constantly also react to a wide range of other stimuli. Innovators, imitators, and laggards will react to different doses. Soft incentives will suffice to activate open-minded market players, while others will only take action when threatened by closure, or insolvency. Adaptive flexibility therefore is crucial to the effectiveness of any formal framework,[1] and a keen understanding of its importance is what sets the evolutionary approach apart from welfare economics and static regulatory policy. In this respect, the jurisprudential responsive regulation approach offers interesting points of contact with evolutionary economics [2, 20].

If we assume the existence of both knowledge deficits and strategic behaviour, adaptation processes and permanent review mechanisms are imperative. Institutions in the market may be able to take over some functions of external institutions, but they will always require some form of supervision involving possibilities of adjustment. Reforming and designing adequate external institutions become permanent economic policy tasks. The question is how policy can shape institutions so that they support the processes of discovery and selection which accompany competition. In the following, a number of assessment criteria will be described and their applicability discussed.[2] They are framed in terms which make them useful in the scientific analysis and assessment of regulatory policies, taking up the methodological challenge to draw specific conclusions from general insights and theories.

3 Assessment Criteria

The primary categories addressed by our analysis are relevance, effectiveness, efficiency, and adaptation flexibility. Relevance means that the planned changes are tested for legitimacy to ascertain whether state intervention is necessary. The criterion of effectiveness allows for analysing targets and ways of reaching them. Efficiency has to do with the costs of regulation and as

[1] For the purpose of this paper, we will define adaptation as the maintenance of functional processes in systems. Adaptation results from cognitive and institutional influences and is not limited to adaptation to the social environment. Flexibility refers to the depth and speed of adaptive processes. A high adaptive flexibility can therefore be characterized as the ability of a system to change quickly and thoroughly so as to maintain the functionality of its processes.

[2] Wegner [32, p. 214] has underscored the desirability of such assessment criteria.

a criterion stimulates the search for lowest-cost combinations; it is discussed in detail below. Adaptive flexibility, one of the key features of evolutionary economics, takes account of learning processes and aims at identifying improved solutions. This rough classification covers the spectrum of assessment possibilities, which is why also the European Commission applies it in policy evaluation. The following section will make it more operational.

Testing legitimacy to ascertain the *relevance* of any specific reform is an approach which can be traced back epistemologically to Immanuel Kant's principle of universalisation and to John Rawls's idea of action taken behind a "veil of ignorance." A legitimacy test for state measures is needed in order to assess the ability of self-regulation to correct deficits and to evaluate corresponding proposals. These require careful evaluation, because regulatory failures might be worse than market failures. Factors included in the scope of this test covers the questions of which specific problem is being addressed, which potential damage costs may be expected, and how great is the political pressure to take action. In this context it also makes sense to pre-assess self-regulation, i.e. to determine whether social groups are able to negotiate solutions, and which mediating function the law or the state should assume in the process. Alternatively, referring to Ronald Coase, assessment would be able to address regulations as to strengthen the legal position of particular groups if their articulation promoted decentralized learning processes, and if no immediate risks had to be averted. The legitimacy test usually privileges institutional reform over and above the establishment of new institutions.

Table 1. List of criteria and questions for the assessment of cross-cutting governance approaches (I) (own compilation, see also acknowledgements)

Criteria		Questions
Relevance	(C1) Process of problem identification, pressure to act	How and by whom is a relevant problem addressed? To what extent does a consensus about causes, effects, and the need to react exist? How urgent is the need for action seen from the actors' perspective? Does the approach address main actors? Is the process stakeholder-driven? Is the process used for priority area identification in line with other stakeholders' agenda? Is it in line with global or regional trends?
	(C2) Decentral solutions, possibilities for compensation	Is there an obvious link to other policy issues, to whom the approach might add negotiated solutions? Does the approach include relevant groups of society? Does it lead to an exchange of (financial or other) resources, which is considered fair and does not lead to additional externalities?

If the results of the legitimacy test are positive, the next step is to assess the effectiveness of institutional reform. Is there a clearly stated *target* with a clear analytical relationship to the specific problem calling for regulation? A verifiable goal is desirable in empirical assessment. Clear criteria for measuring success allow for observing how goals are reached over time, making it possible to downsize an institution as the problems in its remit are solved, and thus to prevent institutional structures from growing ossified and obsolete. Where the target is not clearly defined, effectiveness can nevertheless be assessed by relating a baseline year to a business-as-usual scenario and a scenario of changes effected by regulation. If several targets exist, the relationships among them have to be taken into account – including targets set earlier which entail activities with an impact on the achievement (or not) of new targets. Our assessment approach – as we see it – does not stipulate complete consistency in balancing conflicting goals, though fundamental inconsistencies are to be avoided.

Evolutionary economics takes a sceptical but not hostile view of policy targets. Although Wegner [32, p. 39] suspects that targets "collide with all evolutionary ideas of economic order", it can be argued that evolution, however dynamic, has a direction. Eggertsson [9, p. 1197] similarly supports a process of economic policy development which includes the setting of targets. Meier and Slembeck [16, pp. 84ff., 246ff.] also subscribe to this view. The kind of open development which evolutionary economics call for, hence, depends on a general orientation for which targets are useful.

To combat sceptical views, the assessment of targets also extends to possible measures and potentials for reaching them. Does a given measure propel developments towards the target? Does it, at least, achieve a quantitative deviation from the norm or from a minimum target? Technically speaking, bottom-up analyses determining which solutions are close to market maturity, are needed. Grossekettler [11, p. 548] describes these steps in assessment as the "condition of impulse direction" and the "condition of impulse strength."

The assessment of effectiveness involves two steps for weeding out inappropriate institutional arrangements. The first step excludes obviously ineffective reforms from further consideration. In a second step, the potentials of self-regulation are reassessed. Following the Kaldor-Hicks criterion [35, p. 4ff.], "preferring mutual compensation within social groups rather than central authority's intervention", the possibilities of decentralized control are compared to the risks of state regulation. Does the new governance approach fundamentally limit market processes and does it interfere more strongly with the decision-making power of organizations (e.g. associations) and individuals than is necessary for eliminating market deficits? This second step serves to smooth out obvious snags. The remaining approaches can then be ranked in a provisional order.

Defining a normative criterion is the main problem in *efficiency* assessment. New Institutional Economics is just as sceptical as evolutionary economics when it comes to a static concept of allocation efficiency. In our view,

Table 2. List of criteria and questions for the assessment of cross-cutting governance approaches (II) (own compilation, see also acknowledgements)

Criteria		Questions
Effec-tive-ness	(C3) Targets and strategies	Are there clear and verifiable targets? How consistent are sets of targets in the relevant area beyond the case study? Is the structure suitable for policy deliberations? Does the structure allow for stakeholder participation and interaction on targets and strategies? How consistent is the time horizon of targets with appropriate action? Is there a defined norm or a baseline year?
	(C4) Target implementa-tion	Is there a specific action plan with concrete measures? How can the targets and/or the action plan be related to individual action? Are there performance indicator systems? Are these approaches supported by written and continuously reviewed routines? Do these approaches entail a monitoring of costs (see C5)?

however, this scepticism does not rule out efficiency assessments. Efficiency assessments should address transaction costs, learning processes, and externalities. In our understanding of economic policy as a collective learning process, the dynamic efficiency of coordinated learning processes is more important than static allocation efficiency. In addition to static allocation efficiency, dynamic coordination efficiency also involves a) long-term effects of successive, incremental reforms, b) radical reform (changes in the system), c) generation of new knowledge about solutions which go beyond alleviating situations of asymmetric information, and d) appropriate incentives from economic policy. The concept includes both actor initiatives and reactions from the social environment. Instructive background is to be found in [1, p. 98f.], [23, p. 29ff.], and [17]. The assessment criterion of adaptive flexibility also reflects this concept of efficiency.

An important criterion for assessing the efficiency of new institutions results from the standard function of reducing transaction costs [21, 19]. According to Dixit [7, p. 148], economic policy should take care, at least, to *prevent new or additional transaction costs* when introducing new regulations, its goal being a stable system which reduces the insecurity often attached to interactions. Reducing information asymmetries between social groups is therefore a priority in this context. Measures ensuring that suppliers and consumers have equal access to information and correcting the traditional asymmetry which is so detrimental to small and medium-sized enterprises, will lead to the overall effect of reducing transaction costs in the economy. Admittedly, simply providing access to information is not enough; actors also need support in acquiring knowledge.

A further assessment criterion, the *reduction of negative external costs*, derives from the general function of institutions as constraints. A new regulation which causes additional external costs is to be rejected unless a higher net benefit can be demonstrated. This criterion can also be applied for the reduction of external costs as explicit purpose of testing an existing regulation. In this case, analysis will determine which other external costs would be affected by a change in regulation, and which negative side effects might accompany such a change. Methodologically, the reduction of negative external costs can be assessed through economic analyses as well as empirical studies of the articulation of interest groups. Since the psychology of perceived ownership and loss aversion leads to a disproportional articulation of potential losers, flanking economic analyses are essential.

External costs can "disappear" by shifting them geographically. Economic policymakers and lobbying groups share an interest in shifting burden, which then occasion costs in other parts of the world – an effect which needs to be borne in mind when researching in economic systems and effects. To support longer-term improvements [7, p. 148], analyses of the effects of economic policy must identify where costs are being shifted, and estimate the extent of these costs. On this basis, alternative arrangements can be devised which reduce burden-shifting through collaborative and compensative solutions.

Part of the test should address the "Delaware effect", a term which describes how a reduction of institutional constraints in one area, or state puts pressure on others to follow suit.[3] The phenomenon is also described in terms of a "race to the bottom", or alternatively "race to the top", where positive effects on regions at the forefront of development are discernible [31]. Weak and stagnant systems in developing countries are characterized as "stuck at the bottom" [25] in contrast to the strong and dynamic systems of industrialized countries. The potential pressure to reduce constraints is subsumed in the criterion of external costs, as this yields a logical evaluation. To relax regulations, as we would like to argue, is legitimate as long as new external costs or burden-shifting, are avoided and the change is the result of democratic processes. In such a context, lowering standards can be seen as a sensitive way of exploiting a region's comparative cost advantages. However, if the reduction of certain standards causes new external costs, or shifts them elsewhere, our assessment will arrive at a critical evaluation. Sykes [29, p. 262] distinguishes social and environmental standards, saying that lower social standards gener-

[3] The effects of lax corporate law emerged clearly in the U.S. American state of Delaware. Neighbouring states adapted their regulations in a bid to prevent enterprises from moving away. State regulations on company reporting were tightened at a later date. I would like to thank Bernhard Nagel for communicating information on the situation.

ally do not cause higher external costs,[4] while lower environmental standards always do.

A further assessment criterion based on this principle is the *stimulation of innovation, learning effects, and competition.* New regulations should be oriented on the medium rather than the short term, and aim for improvements beyond the technologies available in the market. Short periods of transition preclude the necessary processes of adaptation and distort competition in favour of a small number of suppliers. The "knowledge-creating competition" model, in contrast, stipulates medium-term periods of transition which enable companies to test a series of hypotheses and develop specific capacities. Whether a new regulation stimulates competition is therefore an important sub-criterion.

A further assessment criterion derived from studies on institutional change evaluates the *desired scale and network effects* of new regulations. These effects occur where a potentially high number of users is interlinked; telephony is an obvious example. Centralized regulation is more likely to meet this criterion than decentralized regulation, so it should be carefully weighed against the advantages of decentralization. According to Sykes [29, p, 259], harmonization generally proves advantageous where preferences are homogenous, and where external effects need to be taken into account. Trachtman [30, p. 337] and Berg [3, p. 461] support this view. Conversely, decentralized solutions are to be preferred where preferences are heterogeneous and externalities can be internalised.

Further assessment criteria on *adaptation flexibility* primarily address learning processes in organizations. Derived from findings on interactions between organizations and institutions [21], these criteria also take up Metcalfe's ideas on adaptive learning in politics [17]. They help to evaluate the institutional risks of "capture of the regulator" by the regulated interest groups and similar processes. The assessment criteria described in the following address issues related to organizations. They are based on the assumption that institutions, if they are to evolve successfully, have to understand the interplay between rules and players of the game, and should react less to changes in relative prices. Following ideas developed in the context of responsive regulation [2, 20], they focus on the activation of third parties which may be expected to have a strong orientation on the common weal (e.g. non-governmental organizations).

One important assessment criteria may be derived from theories according to which institutions have the function of facilitating action [19, 15, 17, 27]. The *principle of free implementation and choice of instruments* should therefore guide the design of institutions, always leaving it to the diverse processes of market implementation to decide which technical solutions and associated services are developed. A "blacklist" of banned instruments is to be preferred

[4] Exceptions are, for example, healthcare and safety at work, where lower standards have a negative external effect on health.

Table 3. List of criteria and questions for the assessment of cross-cutting governance approaches (III) (own compilation, see also acknowledgements)

Criteria		Questions
Efficiency	(C5) Cost reduction	Which internal and external damage costs does the approach try to address? Is there a visible strive for minimizing overall costs? In what ways are transaction costs included? In what ways is there a reduction of external costs? In what ways might new externalities emerge?
	(C6) Positive Side Effects	In what ways does the approach spur incremental or radical innovation? In what ways are processes of diffusion enhanced? Are there tendencies for inertia or is there a systematic effort towards openness for new ideas? What kind of benefits emerge (tangible and non-tangible assets)? To what extent can the approach exploit economies of scale and/or network externalities?
	(C7) Negative Side Effects	Are there systemic leakages, which may lead to problem shifting? Are there incentives for free-riding? Are there new and additional negative externalities?

to a positive list of desired options, as it leaves more possibilities open. The principle of free implementation and choice of instruments should also govern certain markets. For example, although standards of supply may be defined for a national economy or other economic areas, their technical implementation will take different forms and shapes in different regions, taking account of heterogeneous preferences and patterns of demand. In this respect, our approach goes beyond the traditional regulatory principle of allocating an instrument and an agent to each target [11, p. 544ff.]. The reason for this departure lies in our understanding of decentralized learning processes as sources of new strategies for reaching any given target. Specifying the use of at least one instrument would pose unnecessary constraints.

An essential criterion for assessing organizations concerns monitoring of any mechanism. This criterion, however, only applies to regulatory bodies which are, as Karl Popper put it, "properly manned," i.e. have the status of an organization. Examples are authorities supervising capital markets, regulatory commissions and authorities overseeing natural monopolies, as well as international regimes regulating global collective (or even public) goods.[5] "Unmanned" legal institutions can only be judged by their effects and evolve

[5] Standard features of organizations include a secretariat, a conference of parties endowed with decision-making powers, and a number of standing committees in which the parties are represented.

through legislation and legal precedent. Effective monitoring is vital where principal-agent problems arise and regulation is needed to negotiate interests. Monitoring can operate through supervisory boards or similar bodies, budgetary controlling, auditing and accounting, and regulations on reporting. Generally speaking, ex-post monitoring is less problematic than ex-ante regulation.

Assuming that any institutional reform has to preserve sufficient leeway for flexible adaptation, we would like to discuss *evaluation and review mechanisms* as a further assessment criterion. Though the ideal is a framework which obviates the need to introduce process regulations in retrospect, it is only realistic to postulate a certain adaptive flexibility. Recent research on technical institutional change supports this view [18, 19, 15, 21]. Adaptive flexibility allows to allay teething troubles, to tackle new obstacles, and to repair defective framework regulations. Such a mechanism is essential for resolving conflicts of interest where a basic consensus on general principles needs to be reconciled with specific targets or regulations. Institutions for managing conflicting interests, moreover, have to set up formal procedures for resolving conflict. An agent who is given powers and a budget to carry out the institution's tasks can use a general mandate to gradually establish appropriate mechanisms, as Posen [26] has shown in the development of money supply and monetary policy. Héritier [12] discusses this in the context of the European Union. Assessment will have to pay particular attention to the periods of time, procedures and decision-making processes involved in evaluation and review.

A final assessment criterion refers to ideas on deliberation [1, p. 134], i.e. institutional development on the basis of articulation and the deliberation of proposals. *Participation and transparency* are assessed in organizations, with the assessment of participation concerning individuals, organizations and new organizational structures. Formal participation of regulated interest groups harbours risks of ossification and collusion. Appropriate mechanisms of participation anchor an institution in informal rules, and deliberative development takes account of clients' wishes – the mechanism is familiar from stakeholder consultations in companies. Internal participation reduces the risk of individuals being dominated by regulated interest groups. Transparency describes the accessibility of reports and information on individual decisions to outsiders. High transparency exists where the media and representatives of civil society are invited to voice their opinions. This increases the possibilities of articulating the dissent and external knowledge which make up an institution's selection environment.

All in all, the following assessment criteria for institutional reform and design have been identified:

Table 4. Indicative list of criteria and questions for the assessment of cross-cutting governance approaches (own compilation, see also acknowledgements)

Criteria		Questions
Relevance	(C1) Process of problem identification, pressure to act	How and by whom is a relevant problem addressed? To what extent does a consensus about causes, effects, and the need to react exist? How urgent is the need for action seen from the actors' perspective? Does the approach address main actors? Is the process stakeholder-driven? Is the process used for priority area identification in line with other stakeholders' agenda? Is it in line with global or regional trends?
	(C2) Decentral solutions, possibilities for compensation	Is there an obvious link to other policy issues, to whom the approach might add negotiated solutions? Does the approach include relevant groups of society? Does it lead to an exchange of (financial or other) resources, which is considered fair and does not lead to additional externalities?
Effectiveness	(C3) Targets and strategies	Are there clear and verifiable targets? How consistent are sets of targets in the relevant area beyond the case study? Is the structure suitable for policy deliberations? Does the structure allow for stakeholder participation and interaction on targets and strategies? How consistent is the time horizon of targets with appropriate action? Is there a defined norm or a baseline year?
	(C4) Target implementation	Is there a specific action plan with concrete measures? How can the targets and/or the action plan be related to individual action? Are there performance indicator systems? Are these approaches supported by written and continuously reviewed routines? Do these approaches entail a monitoring of costs (see C5)?
Efficiency	(C5) Cost reduction	Which internal and external damage costs does the approach try to address? Is there a visible strive for minimizing overall costs? In what ways are transaction costs included? In what ways is there a reduction of external costs? In what ways might new externalities emerge?
	(C6) Positive side effects	In what ways does the approach spur incremental or radical innovation? In what ways are processes of diffusion enhanced? Are there tendencies for inertia or is there a systematic effort towards openness for new ideas?

Criteria		Questions
		What kind of benefits emerge (tangible and non-tangible assets)? To what extent can the approach exploit economies of scale and/or network externalities?
	(C7) Negative side effects	Are there systemic leakages, which may lead to problem shifting? Are there incentives for free-riding? Are there new and additional negative externalities?
Adaptation flexibility	(C8) Freedom and flexibility	Can relevant actors freely choose among a set of instruments? Is there sufficient flexibility to make investment decisions consistent with the approach's aims? Can actors develop new tools that have an influence on the approach?
	(C9) Evaluation and review	Is there a formal approach for evaluation and/or review? Does it include reviewers outside the approach? Are there clear performance criteria that help to readjust the approach?
	(C10) Participation and transparency	What approaches for participation and transparency exist? Are all relevant groups (affected parties) members of the approach? Do public interest actors hold specific competences? Is the process open for new participants?
	(C11) Control	Which formal and informal control approaches exist? Is there a sufficient division of competences between controlling and controlled actors? What processes ensure independence and power of control over time? What sanctions are foreseen in case of non-compliance?

4 Towards an Application

Each assessment criterion is based on research in its specific area. Combining the criteria enables research to assess different cross-cutting governance approaches in preparation for making decisions. Assessment criteria are helpful in sensitivity analyses, detecting negative side effects, forecasting possible outcomes, and identifying organizational problems. On this basis, empirical research can compare institutions or, more precisely, the effects of institutions for the provision of collective goods. For economic policy, that means increased options for indirect regulation. The systematic exploration of alternatives provided by activities at lower and private levels allows subsidiarity effects.

Methodological studies indicate that there is no single, perfect institution against which all other options fade [19, 27]. According to our approach, no institution is neutral in the sense of being independent of its context of application. The issue an institution is meant to solve and the institutional environment play decisive roles. If we assume limited capacities for assimilating information, economic policy emerges as an adventurous journey of discovery rather than a rational process of optimisation [9, p. 1195]. The task of institutional analysis is, therefore, to systematically compare various second-best solutions and develop rules for solving specific problems. This method corresponds to those concepts of rationality, according to which individuals develop context-dependent decision-making rules and the trend and dynamics of competitive processes derive from the institutional context [15]. Economic policy analysis is thus able to propose successive measures of improvement which move beyond the zero-sum game diagnosed by Stiglitz [28, p. 14] in typical decision-making processes. Such an approach improves the institutional framework's ability to adapt to new situations and can be regarded as an evolutionary version of regulatory policy. The assessment criteria described here also help in implementing ideas on 'process policy' put forward by Wegner [32, p. 156ff., 220ff.].

To reach an overall evaluation of different proposals, however, the assessment criteria need to be specified in more detail. A monetary evaluation of effects is highly difficult to model as there is hardly a sufficient basis of information for calculating probabilities, yielding only approximations which are at best rough estimates. This is especially relevant for cross-cutting approaches. With organizations, on the other hand, monetary evaluation is possible as certain relevant types of cost arise (fixed and variable costs, labour and material costs, investments). In an overall evaluation, a scoring system could be used to compare different institutions, with questionnaires to break down and specify the assessment criteria (see above). Each criterion could be awarded four points determined through ordinal scaling, with a table to illustrate results so that a transparent evaluation of the pros and cons of a specific approach is possible. Fine-tuning and review methods, as used in scientific policy consulting, have proved helpful in this context. In the mid-term, a standardized evaluation matrix for regulatory impact analysis is definitely a possibility.

Table 5. Significance scoring for impact assessments

1 = negative impact compared with the base situation
2 = non-significant impact compared with the base situation
3 = positive impact compared with the base situation
4 = significant positive impact compared with the base situation

Impact assessments are usually confronted with a lack of reliable and homogenous data, which can be characterized as general (the availability of

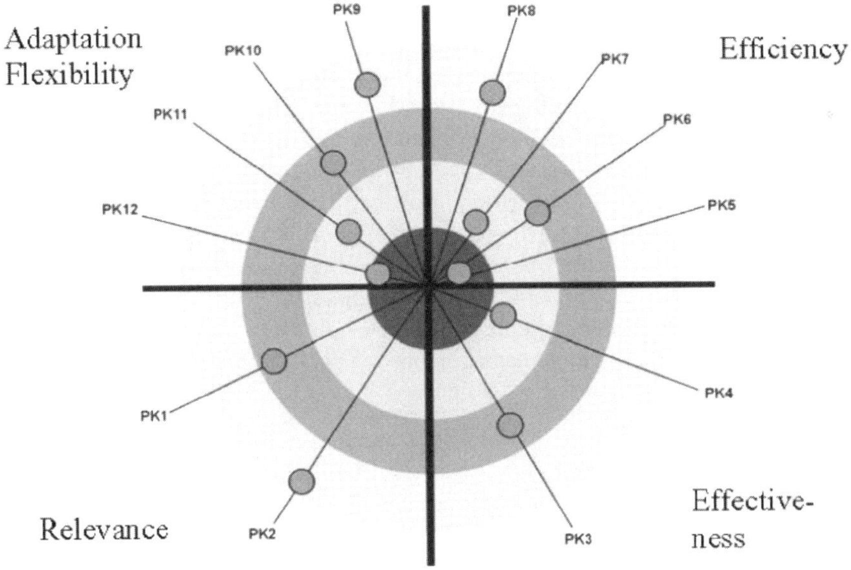

'PK' refers to different criteria, different colours to different scores.

Fig. 2. Visualisation of assessments (draft)

coherent data being the exception). In particular, if private and societal actors are involved, the question of data availability gets crucial. For developing countries data gaps affect almost any field (from economic to social and environmental). Even in the economic field, when some data are available, they are often not reliable because of the importance of informal sectors. Even for developed countries data gaps exist or data are not reliable [5, chap. 5]. Questions to be addressed in empirical studies are:

- Are the available data sufficiently?
- How can the data be compared in order to validate them?
- Can data availability significantly influence the content and validity of the assessment?

5 Conclusions

The present paper uses innovation, institutional, and evolutionary analyses as the basis for setting up assessment criteria which can be applied in scientific policy consulting. Our analysis may therefore be expected to advance in tackling the challenges of dealing with bounded knowledge, involving non-governmental actors, developing and implementing innovations and balancing framework and process regulations. Traditional approaches, such as cost-benefit analysis, have definite methodological deficits in these areas, where, we may conclude, our analysis can complement, or even replace traditional methods of evaluating regulatory impacts and ensuing governance systems. Its field at application is to be found in areas where heterogeneous preferences prevail and technological change offers potential scope for the provision of collective goods, dependent upon flexible cross-cutting governance approaches. It is our claim that these conditions apply to a large number of cases.

The area of application for our approach seems vast. Policy domains like climate, energy, mobility, agriculture, waste, and housing are obvious fields of application because either the complex existing regulation or the new cross-cutting approaches go beyond traditional impact analysis. In these areas, one can expect a mixture of economic incentives, voluntary action, and legal incentives, where more traditional assessments are likely to fall short. Our approach might be useful:

- To pre-select a few instruments from various sources; an econometric modelling of those few approaches can then be done later on.
- To analyse implementation of ongoing processes driven by various actors and, networks where usually unforeseeable side effects emerge; it can be used to review these processes and come up with suggestions for adaptation.

Policy analysis for sustainability will have to rethink individual sectoral approaches and prevailing regulatory tools. Energy, for instance, certainly is important but analysis ought to be combined with impacts on other sectors. Horizontal diffusion will become more important in the next decade of sustainability research. Relaxation on some carefully defined regulations in certain sectors can be legitimate as long as progress in other areas can be achieved. Companies undertaking pioneering efforts in one area or internationally will certainly welcome less pressure from traditional regulation (see e.g. BP demanding tax relief from the UK government). Our paper suggests that well-designed efforts can open up cross-sectoral markets, and also includes some criteria toward a compensation scheme for assessing cross-sectoral approaches. Beyond the areas mentioned, our assessment methodology might also be helpful in areas like provision of public services, economic policy, technology policy, trade policy, and development cooperation.

The present article aims to provide one module in a new form of economic policy consulting which is process-oriented and actively involves actors from the private sector. Policy for sustainability, it turns out, is a collective learning process. Some tentative empirical analysis (see acknowledgements) seems to reveal both the necessity as well as the practicability of our approach.

Acknowledgements

Earlier versions of this paper have been presented at the Seeon conference on innovation and sustainability policy in May 2004 and on the Berlin IHDP conference in December 2004. With fellow researchers at the Wuppertal Institute we tested our assessment criteria in a comprehensive ana-lysis of some framework regulations, regional policy networks and proactive entrepreneurial measures. The focus was on environmentally-oriented reforms (ecological tax reform, demand-side management, promotion of renewable energies, responsible care, 'Eco-Profit', BP emissions trading system, eco-efficiency promoting schemes, Dow Jones Sustainability Index). Our research received generous support from the Japanese Cabinet Office, channelled through ESRI, NRI and MRI, to which we would like to express our gratitude. The study is part of an international programme called 'Millennium Collaboration Projects' (www.esri.go.jp); a first book has been published [5], a second book is in print [6]. I owe thanks to my colleagues Peter Hennicke, Thomas Langrock, Michael Kuhndt, Michael Latsch, Stephan Ramesohl, Holger Wallbaum, Stefan Bringezu, Bettina Bahn-Walkowiak. I also owe thanks to the participants of the "Buchenbach-Workshop on evolutionary economics" in May 2003, especially to Franziska Pamp, Marco Lehmann-Waffenschmidt, and Alexander Ebner.

References

1. Ahrens J. (2002): Governance and Economic Development. A Comparative Institutional Approach. Edward Elgar, Cheltenham Northampton
2. Ayres I. and Braithwaite J. (1992): Responsive Regulation. Transcending the Deregulation Debate. Oxford University Press, Oxford
3. Berg R. vd (2000): Regulatory Competition in Europe. Kyklos 53, 435–466
4. Berkhout F., Leach M. and Scoones I. (eds.)(2003): Negotiating Environmental Change: New Perspectives from Social Science. Edward Elgar, Cheltenham Northampton
5. Bleischwitz R. and Hennicke P. (eds.) (2004): Eco-Efficiency, Regulation and Sustainable Business. Towards a Governance Structure of Sustainable Development. Edward Elgar, Cheltenham Northampton
6. Bleischwitz, R. (ed.) (2007): Corporate Governance of Sustainability: A Co-Evolutionary View on Resource Management. Edward Elgar, Cheltenham Northampton, forthcoming
7. Dixit A.K. (2000): The Making of Economic Policy. A Transaction-Cost Perspective. MIT Press, Cambridge, MA
8. EC = European Commission, DG Enterprise (2004): A Comparison of EU Air Quality Pollution Policies and Legislation with other Countries. Review of the Implications for the Competitiveness of European Industry, Brussels
9. Eggertsson T. (1997): The Old Theory of Economic Policy and the New Institutionalism. World Development 25 (8): 1187–1203
10. Eucken W. (1990 [1952]): Grundsätze der Wirtschaftspolitik. UTB, Tübingen
11. Grossekettler H. (1996): Öffentliche Finanzen. In: Bender D. et al. (eds.): Vahlens Kompendium der Wirtschaftstheorie und Wirtschaftspolitik. Bd. 1, Vahlen, München,: 483–628
12. Héritier A. (ed.)(2002): Common Goods. Reinventing European and International Governance. Lanham et al., Rowman & Littlefield Publ.

13. Jordan A., Wurzel R.K.W. and Zito A.R. (2003): 'New' Instruments of Environmental Governance? National Experiences and Prospects. Environmental Politics 12 (1): 201–224
14. Kerber W. (1997): Wettbewerb als Hypothesentest: Eine evolutorische Konzeption wissenschaffenden Wettbewerbs. In: Delhaes K.v. and Fehl U. (eds.): Dimensionen des Wettbewerbs. Schriften zu Ordnungsfragen der Wirtschaft 52. Stuttgart: 29–78
15. Mantzavinos C. (2001): Individuals, Institutions, and Markets. Cambridge University Press, Cambridge, MA
16. Meier A. and Slembeck T. (1998): Wirtschaftspolitik. Ein kognitiv evolutionärer Ansatz. Oldenbourg, München & Wien
17. Metcalfe J.S. (2001): Institutions and Progress. Industrial and Corporate Change 10 (3): 561–586
18. Nelson R. (2002): The Problem of Market Bias in Modern Capitalist Economies. Industrial and Corporate Change 11 (2): 207–244
19. Nelson R. and Sampat B.N. (2001): Making Sense of Institutions as a Factor Shaping Economic Performance. Journal of Economic Behavior & Organization 44: 31–54
20. Nonet P. and Selznick P. (1978): Law and Society in Transition – Towards Responsive Law. Harper & Row, New York
21. North D.C. (1998): Five Propositions about Institutional Change. In: Knight J. and Sened I. (eds.): Explaining Social Institutions. University of Michigan Press, Ann Arbor: 15–26
22. Olson M. (1982): The Rise and Decline of Nations. Yale University Press, New Haven, London
23. Pelikan P. and Wegner G. (eds.)(2003): The Evolutionary Analysis of Economic Policy. Edward Elgar, Cheltenham Northampton
24. Pierson P. (2000): The Limits of Design: Explaining Institutional Origins and Change. Governance: An International Journal of Policy and Administration 13 (4): 475–499
25. Porter G. (1999): Trade Competition and Pollution Standards: "Race to the Bottom" or "Stuck at the Bottom". Journal of Environment and Development, June: 133–151
26. Posen A.S. (1998): Do Better Institutions Make Better Policy? International Finance 1 (1): 173–205
27. Rodrik D. (2000): Institutions for High-Quality Growth: What They Are and How to Acquire Them. Studies in Comparative International Development 35 (3): 3–31
28. Stiglitz J. (1998): The Private Use of Public Interests: Incentives and Institutions. Journal of Economic Perspectives 12 (2): 3–21
29. Sykes A.O. (2000): Regulatory Competition or Regulatory Harmonization? A Silly Question Journal of International Economic Law 3: 257–264
30. Trachtman J.P. (2000): Regulatory Competition and Regulatory Jurisdiction Journal of International Economic Law 3: 331–348
31. Vogel D. (2000): Environmental Regulation and Economic Integration. Journal of International Economic Law 3: 265–279
32. Wegner G. (1996): Wirtschaftspolitik zwischen Selbst- und Fremdsteuerung – ein neuer Ansatz. Nomos-Verlag, Baden-Baden

33. Witt U. (2003): The Evolutionary Perspective on Economic Policy Making: Does It Make a Difference? Journal of Evolutionary Economics 13/2 pp 77-94 MPI Jena Papers on Economics and Evolution 0101, Jena
34. World Bank (2004): World Development Report: Making Services Work for Poor People. Washington D.C.
35. Zerbe R.O. (2001): Economic Efficiency in Law and Economics. Edward Elgar, Cheltenham Northampton

Comment: Regulatory Choice and Responsive Regulation for Sustainability

Kilian Bizer

Chair for Economic Policy and SME Research, Economics Department, University of Göttingen, Platz der Göttinger Sieben 3, D-37073 Göttingen.
bizer@wiwi.uni-goettingen.de

1 Introduction – The Problem of Regulatory Choice

Economics, as other social sciences, aims to understand social coordination by markets as well as by social institutions. Assumed that markets can fail and that governments may be able to correct such market failure, the crucial question is how to organize the social decision process in order to achieve best results with regard to sustainability. *Raimund Bleischwitz* recommends adopting ten guiding assessment criteria based on four categories. The four categories are legitimacy (or relevance), effectiveness, efficiency, and adaptation flexibility. While it is helpful to organize the principles around these categories, they are also closely intertwined. The purpose of this paper is to show that these linkages can be reflected by adopting the approach of responsive regulation. While efficiency requires instruments which lead to a least-cost solution, adaptation flexibility allows for learning processes of agents as well as administrations. If they learn either to circumvent policy instruments or else to cooperate, policies should adopt to secure effectiveness and efficiency. Moreover, as ineffective policy instruments can never be legitimate, it is obvious that in an ever faster changing social environment the government needs different solutions to the problem of regulatory choice than simply prescribing command-and-control policies, or even merely changing relative prices – both are still important parameters for reform, but both need to be embedded in a broader perspective on regulatory policies which is provided by responsive regulation.

Section 2 of my comment discusses responsive regulation as a regulatory approach as opposed to a purely hierarchical regulation which takes into account the links between *Bleischwitz'* four categories of legitimacy, effectiveness, efficiency, and adaptation flexibility. Section 3 demands a behavioural approach going beyond the traditional *homo oeconomicus* in assessing regulatory policies. The model suggested here is called *homo oeconomicus institutionalis* emphasizing that agents are embedded in institutions which form preferences, perceptions, as well as decision options. Despite remaining well within the rational choice paradigm, a model like this one allows for observations of behaviour, such as instinctive, or emotional behaviour. Section 4 draws some conclusions for the governance of sustainability.

2 Responsive Regulation and the Role of Self-Responsibility

The classic approach to regulation is the intervening state prescribing a policy which clearly indicates allowed and forbidden behaviour. The forbidden behaviour is heavily sanctioned by either fines, prison terms, or even social contempt. In the case of fundamental norms, such as not to kill each other, such an approach is fairly effective, it is relatively easy to organize efficiently, and legitimacy is rarely questioned. As fiction of crime suggests, murders might occasionally be able to set up a "perfect murder", but actual murder cases display very little need to a highly adaptive flexibility as most cases can be sorted out without ever-changing police procedures and new regulatory approaches. Still, is this true for more complex norms as well?

One regulatory field with high relevance to sustainability is the regulation of chemicals as an input in production. Of the approximately 30.000 "old chemicals", little is known about toxicity or other health and environmental impacts. The traditional approach would suggest forbidding all toxic or otherwise harmful substances. Still in doing so the regulatory agency faces three enormous problems: Firstly, it does not know which substances have dangerous impacts, and the information is not only costly, but basically not obtainable without cooperation from the industry. Secondly, impacts may change depending on production processes. And third, some substances may be of high importance in certain industries and impossible to substitute.

In such a context the hierarchical regulation is doomed to fail. Instead, if the state intends to intervene it must adopt an approach which takes into account the incentive situation of the relevant agents and design regulatory options which make it reasonable for them to comply. The Registration, Evaluation and Authorisation of Chemicals (REACh) Directive of the European Commission [5] attempts such an approach by carefully framing responsibilities and demanding information as well as self-responsible policies of firms. Generally speaking, REACh places all major responsibilities with firms rather than administrative bodies. At the same time, this policy is far from being undemanding: Companies marketing substances which fall under REACh must ensure that, along the production chain throughout all intermediate and final users of the substance or the product containing the substance, dangers to health and environment will be reduced. In other words, producers of REACh chemicals must start information and communication processes along the entire production chain to meet the requirements of the directive.

In the regulation of chemicals REACh does not fall short of a paradigm shift [3] from hierarchical regulation to responsive regulation [1, 7]. State intervention is reduced to demand self-responsibility from firms which face rigid liability in case they fail to oblige with the general demands. Liability, as well as the threat of prohibiting individual substances, gives a general incentive for firms to start actions. At the same time, the state does no longer try to collect all relevant information, to process it according to risk assessments, and to design appropriate reactions, but is shifting such responsibilities to firms which are much more likely to have access to such information, which will be able to design appropriate communication and information processes possibly even at lower costs, and have an incentive to reduce risks connected with their substances. The "sticks-and-carrots" approach, characteristic for responsive regulation, is applied by REACh to the regulation of chemicals.

Still, the regulation of chemicals is only one field of application. Most environmental policy fields regulate highly complex decisions and turn more and more towards a responsive approach rather than command-and-control. If sustainability is the overall objective, there hardly is a choice between hierarchical and responsive approaches: As sustainability cannot be specifically defined with regard to individual action, the state must develop general principles of governance allowing individuals to decide for themselves. To ensure that individuals will not merely maximise their own pay-off, responsive regulation demands them to justify their decisions and connect sanctions to non-, or ill-justified decisions.

Self-responsibility plays a vital role in responsive regulation, but it should not be misunderstood as merely shifting the decision to agents and withdrawing [6, 3]. In complex environments, it is impossible for the state to exactly determine obligations of agents. In such circumstances, it is reasonable to impose self-responsibility on agents, making them liable for occurring damages, if they did not fully take into account consequences of their decisions. In order to monitor the results, the state must interact not only with the group of agents directly addressed, but also with other groups which might be affected in cases of negligence on behalf of the norm addresses.

3 Behavioural Models

In order to form expectations of future decisions and to design adequate incentive mechanisms the state must adopt some behavioural model. Such a model should be capable to reflect all relevant behaviour among agents, enabling the state to tailor-cut instrumental approaches for specific groups.

In the currently most favoured model of economic man, agents are assumed to apply utility-maximising behaviour under certain restrictions and based on exogenous preferences. Such a model falls short of allowing a variety of regulatory actions, as agents within such a framework cannot behave habitually, react emotionally, and their preferences do not evolve from within their institutional surrounding. Within the model of economic man, choices can be simply influenced by changing marginal prices. In contrast, in a world of regulatory choices, agents are frequently incapable of calculating and comparing alternatives, use rules of thumb, or take choices regardless of relative prices as long as they can afford to continue as before. The traditional economic approach may prove useful in certain classes of production choices, but it is always in danger to neglect three important aspects:

(1) Decisions in consumption are very often driven by other than purely economic incentives, or, to put it differently, in the narrow perspective of stylized *homo oeconomicus* the relevant preferences are overlooked.
(2) Agents will not perceive all decision options suffering from cognitive limits, rules of thumb, orientation by peer groups, etc.
(3) Finally, market prices definitely play a major role in production and also in consumption decisions, but most decisions are highly complex and much important information might not be adequately reflected in the market price.

Taking into account these aspects, a behavioural model should provide the heuristics for understanding motives of agents. It should account for different modes of behaviour, such as traditional situational utility maximising behaviour (*homo oeconomicus*), as well as for habitual behaviour and emotional behaviour. The model should also reflect that agents often take decisions without knowing the full range of alternative actions, not to mention calculating their respective outcomes. Such cognitive limits occur frequently. In addition, agents can have social preferences. One example is a preference for fair procedures or outcomes.

A model which includes these heuristic supplements of the traditional economic model can be called *homo oeconomicus institutionalis*. Again, the model is taken as heuristics to systematize behaviour and to design additional and new incentives in order to change behaviour. This broader approach to behaviour provides also a broader approach to the applicable instruments which cannot be based on relative prices alone, even though relative prices constitute the major source of behavioural change.

An important difference between the institutional approach to behaviour and the traditional economic approach is to ask for institutional barriers of new incentives. For example, along the production chain knowledge about hazardous substances within the product may not be fully perceived as is the case for varnishers applying paints containing volatile organic compounds. Even though basic information concerning characteristics of the content and health risks is displayed on containers, varnishers might not follow instructions due to underestimating the risk. Such cognitive limits can turn financial incentives ineffective if they are not accompanied by additional informational and educational instructions.

At the same time, new economic incentives, such as charges, frequently bring about a signalling effect as awareness is drawn to a certain issue. Even though agents might not be able to actually calculate the costs and benefits

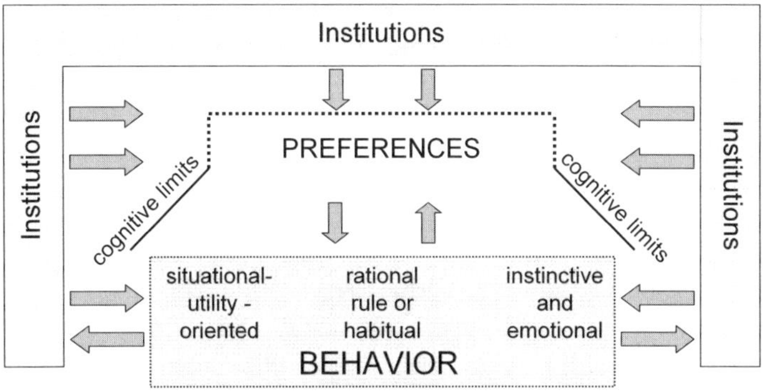

Fig. 1. The behavioural model of *homo oeconomicus institutionalis*
Source: based on [2]

of alternatives, the new economic incentive causes them to reconsider their decisions and to look for alternatives. Such a signalling effect can well be used to trigger changes in behaviour. The approach of *homo oeconomicus institutionalis* reflects that a policy mix of command-and-control, economic incentives, voluntary contributions as well as informational support policies might bring about the best results within the context of responsive regulation as actual behaviour of agents is taken into consideration forming the policy approach.

4 Conclusion

Environments which become more complex make it difficult, if not impossible, for the state to intervene with precise obligations fitting each relevant agent. Equally important is that general objectives, such as sustainability, can hardly be tailor-cut to the behavioural options of certain agents. The state today simply does not know what option of the agent is more sustainable tomorrow. The shift away from clear command-and-control policies is therefore reasonable, but requires new guidance principles. *Raimund Bleischwitz* supplies such guidance by formulating ten assessment criteria based on the four categories legitimacy, effectiveness, and efficiency as well as adaptation flexibility. All four categories deserve full attention of the policy makers as they are closely intertwined. *Bleischwitz'* recommendation to consider the general approach of responsive regulation highlights these linkages, as it takes into account the incentive situation of the addressees of policies. The more differentiated the policy approach, the more it is necessary to provide legitimacy, to ensure effectiveness throughout implementation, and to allow for efficiency and flexibility towards changing behaviour as well as social and environmental conditions. In order to do so, an equally differentiated model of behaviour must be adopted to reflect the many characteristics of behaviour within the policy instruments applied in the approach of responsive regulation. Responsive regulation is less a form of government, but rather of governance as it includes the interactions between groups of agents and the state. Within such an interactive framework crucial sustainability issues, such as the regulation of chemicals, must be addressed in the future.

References

1. Ayres I. and Braithwaite J. (1992): Responsive Regulation. Transcending the Deregulation Debate, Oxford University Press, Oxford
2. Bizer K., Hüttig C. und Führ M. (Hrsg.) (2002): Responsive Regulierung – Beiträge zur interdisziplinären Institutionenanalyse und Gesetzesfolgenabschätzung, Mohr & Siebeck Tübingen
3. Bizer K. and Führ M. (2007): REACh as a Paradigm Shift in Chemical Policy - Responsive Regulation and Behavioural Models, Journal of Cleaner Production (JCLP), 327-334, Vol. 15, 4
4. European Commission (2002): European Governance – a white paper. COM 2002, 428 final
5. European Commission (2003): Regulation, Evaluation and Authorization of Chemicals (REACH). COM 2003, 644 final
6. Führ M. (2003): Eigenverantwortung im Rechtsstaat, Duncker & Humblot, Berlin
7. Nonet P. and Selznick P. (1978): Law and Society in Transition – Towards Responsive Law. Harper & Row, New York

Printing: Krips bv, Meppel
Binding: Stürtz, Würzburg